THE SPACE SHUTTLE OPERATOR'S MANUAL

THE SPACE SHUTTLE OPERATOR'S MANUAL

KERRY MARK JOELS · GREGORY P. KENNEDY

Designed by
DAVID LARKIN

BALLANTINE BOOKS   NEW YORK

Published in the United States by Ballantine Books,

a division of Random House, Inc., New York, and simultaneously in Canada

by Random House of Canada Limited, Toronto, Canada.

Library of Congress Catalog Card Number: 82-90220

ISBN 0-345-30321-0 (trade paper)

ISBN 0-345-30751-8 (hardcover)

This edition published simultaneously in trade paperback and Del Rey hardcover.

Manufactured in the United States of America

First Edition: October 1982

10 9 8 7 6 5 4 3 2 1

Design © David Larkin, Becontree Press

We are grateful to the staff at
Lyndon B. Johnson Space Centre, Houston, Texas
and to Rockwell International, North American Space Operations
for their help in the preparation of this volume.

CONTENTS

To: New Astronauts

Welcome aboard! This manual is your guide to your flight. It contains information on Space Shuttle systems and flight procedures. The book is arranged to follow the progress of your flight from boarding through landing. Keep this manual nearby throughout your mission—the checklists and instructions it contains make it a quick reference guide while flying the Shuttle.

Congratulations! We hope your flight is rewarding and fascinating.

Sincerely,
DIRECTORATE FOR CREW TRAINING

This first chapter provides a brief, generalized explanation of what the Space Shuttle is and how you reach orbit. The chapters following it will provide specific detail on the various capabilities and phases of flying this craft.

SOLID ROCKET BOOSTER   12 FT. (3.7M) DIAMETER

EXTERNAL TANK   27.5 FT. (8.4M) DIAMETER

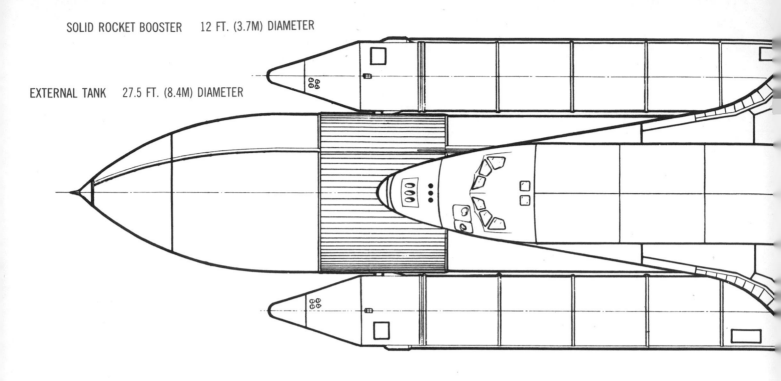

WEIGHTS

ORBITER (EMPTY)	165,000 POUNDS (75,000 KG.)
EXTERNAL TANK (EMPTY)	78,100 POUNDS (35,500 KG.)
SOLID ROCKET BOOSTER (EMPTY)	185,000 POUNDS (84,100 KG.)
ENTIRE VEHICLE, FUELED FOR LAUNCH,	
	4.4 MILLION POUNDS (2 MILLION KGS)

149 FT. (45M)

154 FT. (47M)

184 FT. (56M)

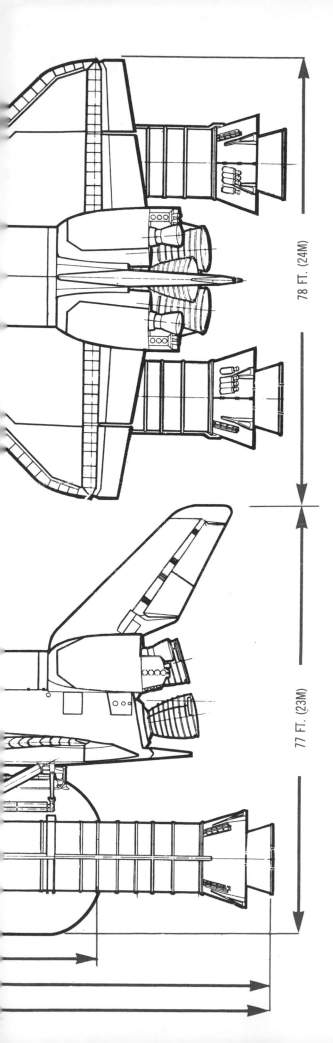

78 FT. (24M)

77 FT. (23M)

How much does it cost to launch a satellite? Before the Space Shuttle, all launch vehicles were "expendable"; that is, they were used only once. In mid-1981, an expendable Delta rocket cost about $25 million. For that price, you could place about 5,000 pounds (2,300 kilograms) into orbit. The reusable Space Shuttle costs about $35 million per flight and carries 65,000 pounds (29,500 kilograms) of payload. In other words, the Shuttle can lift 13 times as much as a Delta at only 1½ times the cost.

The Space Shuttle is an attempt to make outer space easier to reach. It has often been compared to an airliner, but this isn't quite right. First of all, it wasn't designed to carry people. Second, even though Shuttle flights occur frequently, they are anything but routine. Instead, the Shuttle can better be described as a "space truck" designed to carry cargo—satellites—into space.

Three main segments make up the Space Shuttle: a reusable Orbiter, a pair of solid-propellant boosters, and a large liquid-propellant tank.

The Orbiter is the principal part of the Shuttle. Designed to last for one hundred flights, this winged vehicle is part spacecraft, part aircraft. It carries payloads in a 15-by 60-foot (4.5-by-18-meter) cargo bay that dominates the middle of its fuselage. Forward of the cargo bay, you find the crew compartment where you live while in space. Behind the cargo bay, in the aft fuselage, are three engines that propel the Orbiter during launch. Reusable insulation on the craft's exterior enables it to survive the searing heat of atmospheric entry. After entry, the vehicle becomes a glider as it makes an unpowered approach and landing on a runway. On-board auxiliary power units provide power for the hydraulic system that operates the control surfaces during atmospheric flight. Hydrogen and oxygen are combined in fuel cells to generate electricity and water.

The three engines on the base of the Orbiter, called the Space Shuttle Main Engines (SSMEs), are the most advanced liquid-fuel rocket engines ever built. They burn liquid hydrogen and liquid oxygen under high pressure to create as much thrust as possible. Each SSME has a rated thrust of 375,000 pounds (1.6 million newtons) at sea level. The thrust can be varied from 65% to 109% of this rated value.

Propellants for the SSMEs are contained in a large tank, called, appropriately, the External Tank (ET). The tank, made of aluminum, is 154 feet (47 meters) long and 27.5 feet (8 meters) in diameter. It is the largest single Space Shuttle component and the only part not reused. When empty, it is allowed to enter the atmosphere to break apart and burn up over the Indian Ocean. The ET attaches to the belly of the Orbiter. Two large conduits feed propellants into the Orbiter's aft fuselage.

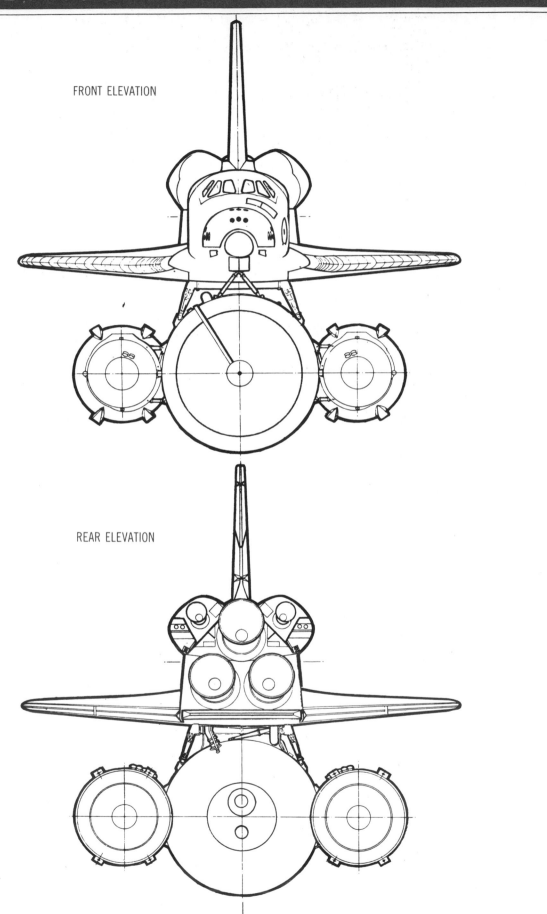

FRONT ELEVATION

REAR ELEVATION

At launch, two large solid-propellant rockets are attached to the tank, one on each side. These are the Solid Rocket Boosters (SRBs), which provide most of the power to lift the Shuttle off the pad and propel it during the first two minutes of flight. After their propellants are consumed, the empty boosters separate from the Shuttle. They descend on parachutes and land in the ocean. The empty motor casings are recovered and reused. These are the largest solid-propellant rockets ever flown—149 feet (45 meters) long and 12.4 feet (3.5 meters) in diameter—and they are the first ever designed for reuse.

The Space Shuttle, an amazing flying machine, provides a means for reaching space, but it cannot open the "space frontier" by itself. The Space Transportation System (STS) has been devised to support **all** types of space missions. The Space Shuttle is the backbone of this system, for it is the conveyance "truck" that gets things there. Accessories for the Shuttle complete the flying portion of the STS.

Most communications satellites are positioned 22,300 miles (35,900 kilometers) above the earth, where it takes 24 hours to complete one orbit (see 3.7). However, the Orbiter, the central part of the Space Shuttle, is limited to orbits below 690 miles (1,100 kilometers). Once the Shuttle delivers a satellite to a low orbit, a rocket motor attached to the satellite can boost it to the desired higher orbit. Such rocket motors are built as part of the STS so that the Shuttle can launch these satellites.

In addition, the STS also includes ground facilities where the Shuttle is assembled, launched, and recovered. Together, all portions of the Space Transportation System provide an economical, efficient way to launch all types of satellites—scientific, application, planetary, or military—into space.

Finally, there is one more task the Shuttle can perform: it can retrieve satellites already in orbit. This means that a satellite that breaks down can be returned to earth for repair. This is the first time we've been able to do such work and the capability is unique to the Shuttle.

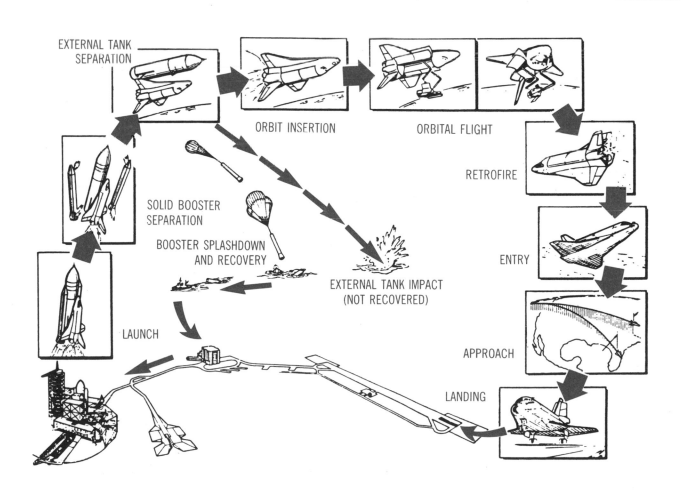

EXTERNAL TANK SEPARATION

ORBIT INSERTION

ORBITAL FLIGHT

RETROFIRE

ENTRY

APPROACH

LANDING

SOLID BOOSTER SEPARATION

BOOSTER SPLASHDOWN AND RECOVERY

EXTERNAL TANK IMPACT (NOT RECOVERED)

LAUNCH

Rocket, spacecraft, aircraft—the Space Shuttle is all of these as it flies through launch, orbit, and landing. This section explains how it performs these functions on a typical flight. The events described reflect a normal sequence; for off-normal conditions and emergency procedures, see Section 4.

Most prelaunch checks are computer controlled, but a special crew boards the Orbiter at T–5 (T minus 5) hours to make sure all switches are set correctly. You board the Orbiter about 2 hours before lift-off. The hours pass quickly because you are kept busy making sure everything is ready for flight.

At T–3.8 seconds the computers command the three main engines to fire. The first starts at T–3.46 seconds, followed by the other two at 120-millisecond intervals. You hear the engines and feel a forward lurch as the Shuttle "twangs" about 40 inches (1 meter) in the direction of the tank.

The engines build up to 90% thrust by T–0, and a timer set for 2.64 seconds starts. When it reaches zero, the Solid Rocket Boosters ignite. The 2.64-second delay allows the Shuttle to be upright at solid-booster ignition as it sways from the twang.

Thrust from the Solid Rocket Boosters builds up quickly and the vehicle lifts off at T+3.0 seconds. From the upper flight deck, you can see the launch tower drop from view. Eight seconds after lift-off the Shuttle rolls 120° to the right, so you will be "heads down" as you climb and arc out over the ocean. During ascent the thrust of the solid motors decreases and the main engines are throttled to keep acceleration below 3 g (3 times earth gravity).

About 50 seconds into the flight the Shuttle reaches the speed of sound, Mach 1. By the time the solid motors consume their propellants (T+2 minutes and 12 seconds) you have reached Mach 4.5 and an altitude of 28 miles (45 kilometers). The empty motor casings separate from the vehicle and land in the ocean via parachute. They are retrieved and later reused.

The Shuttle now consists of the Orbiter and External Tank. It continues to gain speed and altitude; 6½ minutes into the flight, you are traveling 15 times the speed of sound at an altitude of 80 miles (130 kilometers). Flying a path resembling a roller coaster, the Shuttle begins a long shallow **dive** to 72 miles (120 kilometers). During this maneuver, you experience the

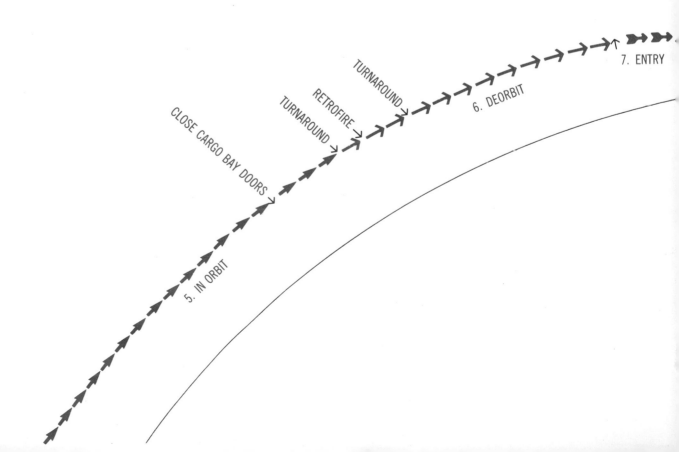

maximum acceleration of 3 g. Near the end of the dive, 8½ minutes after you left the ground, the **main engine cut-off** (MECO) command is given. The External Tank is discarded 20 seconds later. The Orbiter maneuvers down and to the left of the tank which will splash down in a remote ocean area. Remember—throughout the ascent, you travel "upside down" with your head toward the ground. The Orbiter will not assume an "upright" position until you are in orbit—and then it doesn't matter, since you are weightless.

Following ET separation, the **orbital maneuvering system** (OMS), consisting of two 6,000-pound-thrust (26,700-newton) engines, fires to place the vehicle in a low elliptical orbit. Half an orbit later, about 45 minutes after launch, the OMS engines fire again to place you in a higher circular orbit, generally 250 miles (400 kilometers) high.

One of your first tasks after reaching orbit will be opening the cargo bay doors. This is vital—the radiators that shed excess heat generated by the Orbiter are on the inner surfaces of the doors. If the doors remain closed, heat builds up within the vehicle and the mission will have to be aborted within 8 hours.

Once the doors are open and all onboard systems are checked, the mission can proceed. An average mission lasts 7 days, although flights of up to 30 days are possible.

Near the end of the flight, you close the cargo bay doors and the Orbiter turns around so it flies tail first. About 60 minutes before landing (L–60 minutes), the OMS engines fire to slow the vehicle so it will enter the atmosphere. A pitch maneuver turns the Orbiter around with its nose pointed forward and up at an angle between 28° and 38°. Atmospheric entry begins at an altitude of 400,000 feet (122,000 meters) and a distance of about 5,000 miles (8,000 kilometers) from the landing site.

During entry, the tremendous energy of the Shuttle (which begins entry at a speed of 16,500 mph) is dissipated by atmospheric drag. This generates a great deal of heat—portions of the Orbiter's exterior reach 1510°C (2,750°F). The heat ionizes air molecules near the spacecraft, so from about L–25 minutes to L–12 minutes the vehicle is enveloped in a sheath of electrically charged particles that blocks radio signals between the spacecraft and the ground. This means you will be out of communication during one of the most critical portions

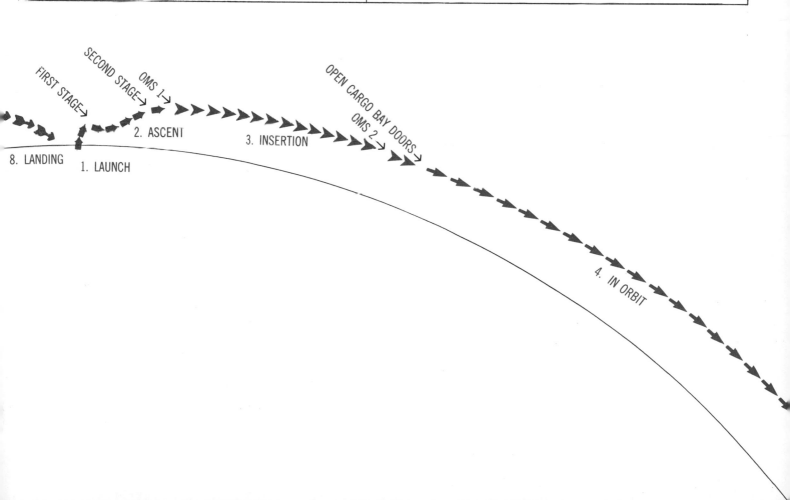

of the flight. Don't worry though—entry maneuvers can be flown by either the onboard computers or the pilot without ground communication.

Rate of descent and range are controlled by bank angles; the steeper the bank angle, the greater the descent rate (and, by the way, the greater the entry heating rate). The Shuttle's nose swings left and right as several hypersonic banks are made to control the flight path. At around L–20 minutes, maximum heating occurs.

As the craft descends, it becomes a glider—an extremely large, heavy one. The transition from control by spacecraft reaction control jets to aerodynamic-control surfaces is made in stages. When entry begins, the forward maneuvering jets are inhibited. The aft jets are used until a dynamic pressure of 10 pounds per square inch (psi) (69 kilopascals) is recorded, which is when the ailerons can be used and the aft roll jets are turned off. At a dynamic pressure of 20 psi (138 kilopascals), the aft pitch jets are deactivated as the elevators become

effective. Finally, at a speed of Mach 3.5 and an altitude of 45,000 feet (13,700 meters) the rudder becomes effective and the aft yaw jets are shut down.

Descending at a rate of 10,000 feet (3,000 meters) per minute, your 100-ton (90,000-kilogram) craft is on a glideslope about 7 times greater than that of a commercial airliner—20°.

At an altitude of 1,750 feet (530 meters), the nose is pulled up in a "preflare" maneuver to reduce your glideslope to 1.5°. By the time this maneuver is complete, you are only 135 feet (41 meters) above the ground.

The landing gear do not extend until the Orbiter is below 90 feet (27 meters), 14 seconds before landing. After the gear extend and lock, a final flare is made to reduce your airspeed to 215 mph (345 kph). The Orbiter lands first on its main gear then drops onto the nose wheels. It rolls to a stop and is met by a convoy of ground servicing vehicles. Your flight is over.

So far, five Orbiters have been named. They are:

ORBITER VEHICLE NUMBER	NAME	REMARKS
OV-101	**ENTERPRISE**	Built as a test vehicle and used for landing tests in 1977. This vehicle not intended for space flight.
OV-102	**COLUMBIA**	Second Orbiter built and first to fly in space on April 12, 1981.
OV-099	**CHALLENGER**	Original airframe built as a test article; modified and finished as a flight vehicle.
OV-103	**DISCOVERY**	Named after Hudson's ship, which in 1610–11 searched for Northwest Passage between Atlantic and Pacific Oceans, and Captain Cook's ship when he discovered Hawaii.
OV-104	**ATLANTIS**	Named for Woods Hole Oceanographic Institute research ship used from 1930 to 1966.

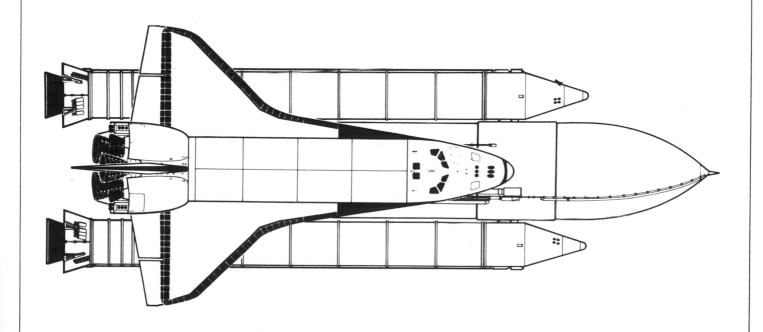

Five onboard computers handle all data processing for your flight. These computers can operate independently, or together. They check on each other, and vote when necessary to solve an argument. Two magnetic-tape **mass memories** providing 34 million types of information add to the total memory of the computers themselves. This data-processing system also has **multiplex-demultiplexers** to translate signals to and from the Orbiter's systems and sensors into computer language, and various displays to show you what is happening and let you talk to the system. These displays include **cathode-ray tubes** (CRTs), **electronic displays**, and **keyboards** to tap in (enter) commands or ask questions. Although the keyboard has only thirty-two buttons, it lets you ask over one thousand questions.

The program for your flight is broken into nine major parts, each with several operational sequences. One sequence controls each flight phase. The programs are written in the **high-order assembly language/shuttle**, or HAL/S.

The sequence program codes are as follows:

Program Code	Mission Phase
901	Preflight Monitor
101	Prelaunch; T–20 minutes to launch
102	Launch to solid booster separation
103	Solid booster separation to main engine cut-off
104	Main engine cut-off through orbital maneuvering system burn #1
105	Orbital Maneuvering System burn #1 and burn #2
106	Orbital insertion and coast
201	On-orbit coast
202	In-flight maneuvering
801	Orbital check-out

Program Code	Mission Phase
301	Preparation for retro-fire (deorbit burn)
302	Retro-fire and preentry coast
303	Preentry system check and monitor
304	Atmospheric entry
305	Atmospheric flight and landing
601	Return-to-launch-site abort
602	Glide option #1 for return-to-launch-site abort
603	Glide option #2 for return-to-launch-site abort

These are general codes; specific ones for your flight are in your flight-data file.

The basic crew consists of three people—commander, pilot, and mission specialist. Overall crew safety and flight execution is the responsibility of the commander. The pilot, as second in command, assists the commander. The mission specialist coordinates payload operations and carries out mission scientific objectives.

Additionally, up to four "payload specialists" may fly on a given mission. Payload specialists are not professional astronauts; rather, they are scientists, engineers, or physicians selected by the organizations that built the payloads they operate. They may be trained to operate other payloads and will be trained in Space Shuttle housekeeping. All payload specialists must also know normal and emergency flight procedures.

Whether you are a commander, pilot, mission specialist, or payload specialist, your living quarters are in the Orbiter's **crew compartment**, which has three levels or **decks**. The **flight deck** is on top. It contains the flight controls and crew stations for launch, orbit, and landing. Below the flight deck, in the **mid-deck**, you will find the accommodations for eating, sleeping, hygiene, and waste disposal. The bottom deck, called the **equipment bay**, is just that—an area containing parts of the waste disposal and life support systems as well as an equipment-storage area.

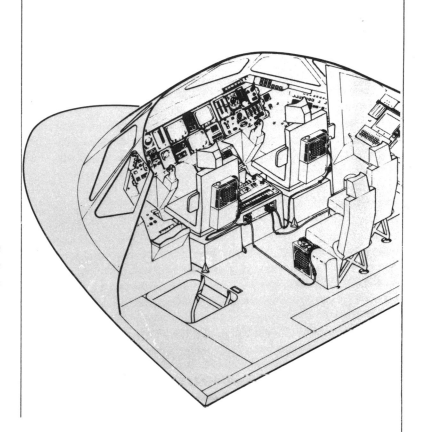

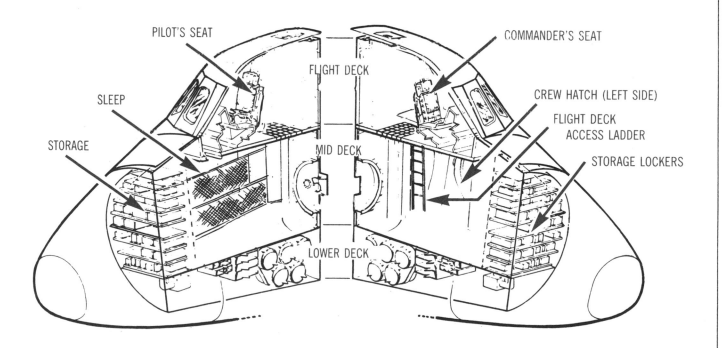

PILOT'S SEAT

COMMANDER'S SEAT

FLIGHT DECK

SLEEP

CREW HATCH (LEFT SIDE)

FLIGHT DECK
ACCESS LADDER

STORAGE

MID DECK

STORAGE LOCKERS

LOWER DECK

FLIGHT-DECK SEATING

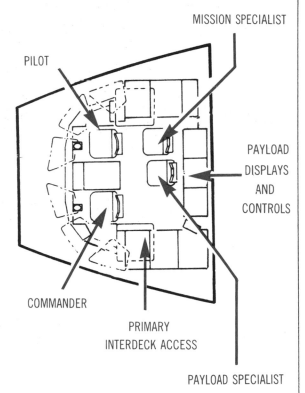

MISSION SPECIALIST

PILOT

PAYLOAD DISPLAYS AND CONTROLS

COMMANDER

PRIMARY INTERDECK ACCESS

PAYLOAD SPECIALIST

MID DECK

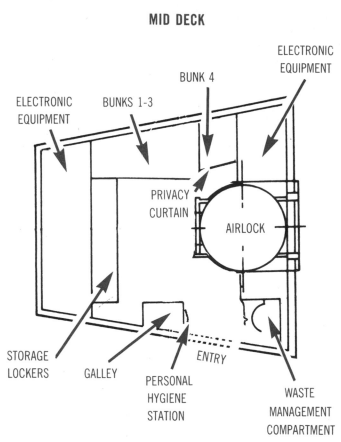

ELECTRONIC EQUIPMENT

BUNKS 1-3

BUNK 4

ELECTRONIC EQUIPMENT

PRIVACY CURTAIN

AIRLOCK

STORAGE LOCKERS

GALLEY

PERSONAL HYGIENE STATION

ENTRY

WASTE MANAGEMENT COMPARTMENT

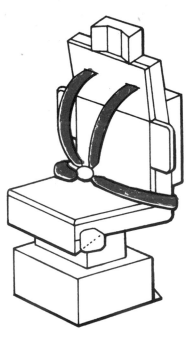

COMMANDER/PILOT'S SEATS

SPECIALIST SEATS

Crew Compartment—Storage

Most of your small equipment is stored in **lockers** on the forward bulkhead of the mid-deck. The contents of each locker are listed on its door.

To open a locker, flip up the thumb latches on the two upper corners and turn the latches until they open. Open the door and pull out the drawer. Each drawer has foam-rubber spacers to hold its contents in place. This prevents small objects from floating out when you open a drawer.

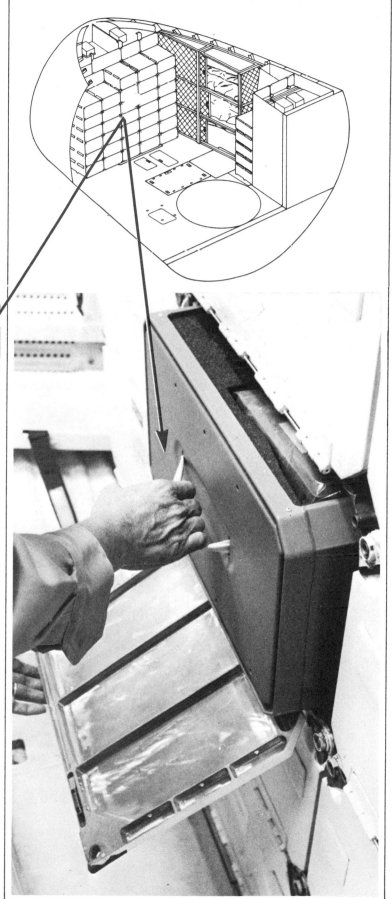

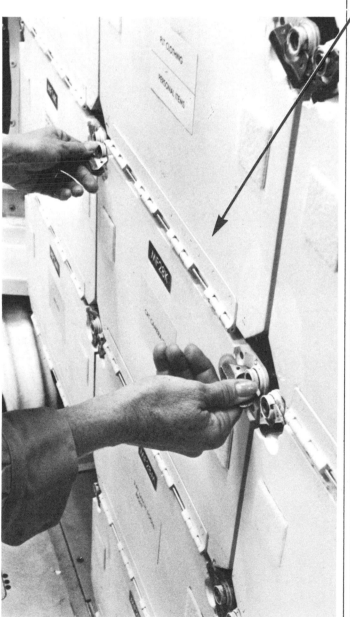

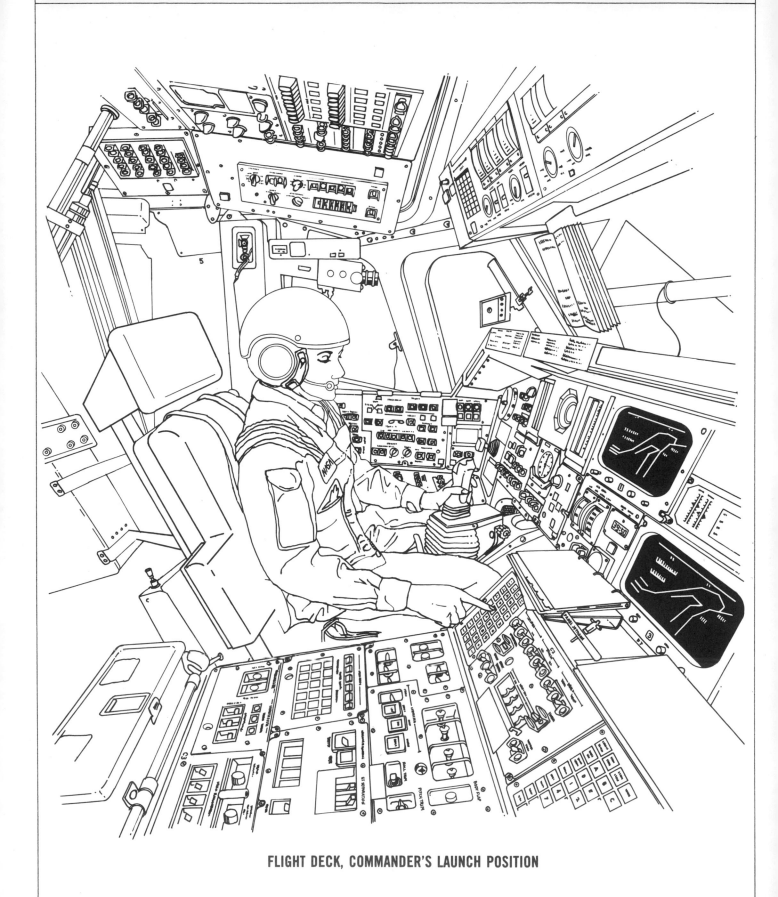

FLIGHT DECK, COMMANDER'S LAUNCH POSITION

PILOT'S LAUNCH POSITION

You and the rest of the crew board the Shuttle about two hours before launch. By the time you enter the spacecraft, the handover/ingress team will have already spent several hours in the Shuttle making sure everything is ready for your flight.

Each mission is different, so the exact sequence and timing of countdown events varies slightly. However, a simplified version of the launch-and-ascent checklist follows for your use. You can use the checklist and the instrument panel drawing to "rehearse" your flight and

as a reference during the flight. This key should help you locate the controls and displays:

Panel	Location
F	Front
L	Left side, next to commander
C	Center console
R	Right side, next to pilot
O	Overhead
CRT	Cathode-ray tube (TV screen)

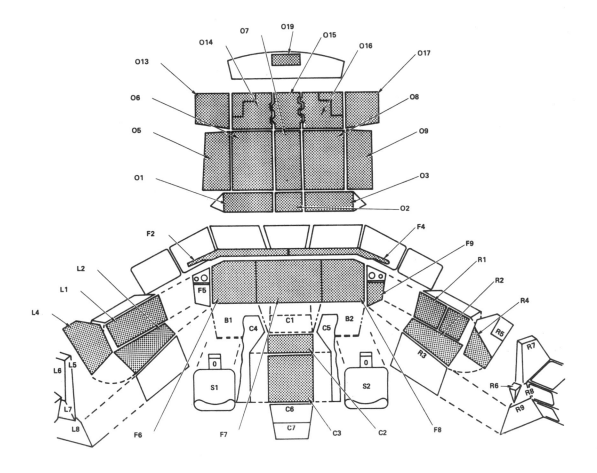

FLIGHT DECK

Checklist

Most countdown events are automatically controlled, or require no action from you; they're marked with an asterisk (*). But there is still plenty for you to do. As you would expect, the tempo picks up as you get closer to lift-off and you will be quite busy in the last nine minutes before launch.

The dialog that accompanies the checklist will

indicate appropriate spoken instructions between the cockpit and the ground controllers. The actual dialog will vary from crew to crew and even from person to person within the crew, but the dialog in this section will give you a reasonable guideline to what you will say and hear. You will use the name of the spacecraft you actually fly, but for this mission we will use **Columbia**.

You enter the crew compartment through a 40-inch (1-meter)-diameter circular hatch on the Orbiter's left side. This hatch opens outward and leads directly into the mid-deck. Since an Orbiter poised for launch is upright (i.e., standing on its aft end with its nose pointing straight up), the mid-deck back wall becomes your floor.

The commander, pilot, mission specialist, and one payload specialist sit in the flight deck for launch and landing. Seats for the other payload specialists are in the mid-deck. Once you crawl into the appropriate deck, footrests and handholds are provided to help you crawl into your seat. When you are strapped in, ready for launch, you are on your back in a sitting position. This posture best enables you to withstand the acceleration you experience during lift-off.

After you reach orbit, the seats for the mission and payload specialists are folded and stowed until the end of your flight. Don't worry about not having enough seats while you're in orbit; in weightlessness, a chair is an unnecessary piece of furniture.

TIME	EVENT	PANEL	PROCEDURE	VOICE 1.18
(Takeoff minus hr:min:sec)				(Y = You; G = Ground controller)
*T−5:00:00	Begin final countdown.			
*T−4:30:00	Begin filling liquid-oxygen tank in External Tank.			
*T−2:50:00	Begin filling liquid-hydrogen tank in External Tank.			
T−1:50:00	Enter Orbiter.		Enter through side hatch, climb into seats.	
T−1:35:00	Unstow cue cards.		Velcro-backed cards are removed from flight-data file and attached to the instrument panel.	
T−1:30:00	Communications check with Launch Control.	05, 09	Launch control will contact you on air-to-ground channel 2, both intercom channels, and the air-to-air channel. Communication controls are on panels 05 and 09, the commander's and pilot's overhead panels. Select proper channel.	G: *Columbia*, this is Launch Control. Radio check, over. Y: Roger, out.
T−1:25:00	Communication check with Mission Control.	05, 09	Check both air-to-ground channels in the same way you did for the launch control check.	G: *Columbia*, this is Mission Control. Y: Roger, out.
T−1:20:00	Abort advisory check.	F6	Launch control will light the abort light on the front instrument panel; bright, dim, then off.	G: *Columbia*, this is Launch Control. Ready abort advisory check. Y: Roger, check is satisfactory, out.
T−1:10:00	Side hatch closure.		The ground crew closes and secures the side hatch.	G: *Columbia*, this is Control. Side hatch is secure. Y: Roger, we copy.

TIME	EVENT	PANEL	PROCEDURE	VOICE 1.19	
T−1:05:00	Cabin leak check.	L2, 01	Close both cabin vent switches; monitor cabin-pressure gauge above commander—it will increase to 16.7 psi.	Y:	Control, this is *Columbia*. We show cabin pressure nominal (normal), over.
				G:	Roger, out.
T−0:51:00	Inertial measurement unit (IMU) preflight alignment.	CRT 2	Watch CRT #2; it should report IMU stabilized for launch position of 28 degrees, 36 minutes, 30.32 seconds north latitude and 80 degrees, 36 minutes, 14.88 seconds west longitude (location of launch pad 39A at Kennedy Space Center).	Y:	Control, this is *Columbia*. IMU alignment complete. We show two-eight degrees, three-six minutes, three-zero point three-two seconds north, by eight-zero degrees, three-six minutes one-four point eight-eight seconds west. Over.
				G:	Roger, *Columbia*, out.
T−0:50:00	Water-boiler preactivation.	R2	BOILER CNTLR (boiler controller) switches (all three)—ON. BOILER N$_2$ SPLY (nitrogen supply) switches (all three)—ON.	Y:	Control, this is *Columbia*. Boiler control switch—on. Nitrogen supply switch on, over.
				G:	Roger, out.
T−0:32:00	Primary avionics software system/backup flight system (BFS) transfer preparation.	06	GPC MODE 5 (general purpose computer, mode 5) STBY (stand-by) light will illuminate.	Y:	Control, this is *Columbia*. GPC, BFS complete, over.
				G:	Roger, out.
		C3	BFS CRT (backup flight system display) select 3 + 1 position BFS CRT DISPLAY—ON CRT screen #3 will display the backup flight system guidance navigation and control memory.		
		C2	On computer keyboard enter: ITEM 2 5 EXEC (execute).		
*T−0:30:00	Ground crew secures white room and retires to fall back area.			G:	*Columbia*, this is Control. Ground crew is secure, over.
				Y:	Roger, out.

TIME	EVENT	PANEL	PROCEDURE	VOICE
T–0:30:00	Orbital maneuvering system (OMS) pressurization.	C3	OMS ENG switches (both)—ARM/PRESS.	
		01	Check cabin pressure—it should be 16.7 psi, Cabin leak check is complete.	
	Cabin vent.	L2	CABIN VENT switches (both)—OP (open). Klaxon alarm should sound.	Y: Control, this is *Columbia*. OMS pressure on . . . Cabin vent complete, over.
		01	CABIN PRESS—it should decrease.	G: Roger, we see that, out.
T–0:25:00	Voice check.	05, 09	Commander and pilot simultaneously conduct voice checks with Mission Control center. Landing weather data for return-to-launch-site abort or abort once-around is updated.	Y: Control, this is *Columbia*. Commander's voice check, over. G: Roger, out. Pilot: Control, this is the pilot; voice check, over. G: Roger, out.
T–0:21:00	Close vent valves.	L2	CABIN VENT switches (both)—CL (closed).	
T–0:20:00	10-minute hold (Countdown clocks are stopped to allow catch-up on any behind schedule activities).			
T–0:20:00	Load flight plan OPS-1 into computer.	C2	Flight program is loaded into computers. ERR LOG switch to RESET; CRT #1 will indicate any guidance navigation and control system faults. On computer keyboard, enter: SPEC 9 9 PRO. CRT #2 will indicate your launch trajectory.	Y: Control this is *Columbia*. Flight plan is loaded into the computer, over. G: Roger, out.
T–0:19:00	Load flight plan OPS-1 into BFS.	C2	On computer keyboard, enter: SPEC 9 9 PRO.	
		06	GPC MODE 5—RUN.	

TIME	EVENT	PANEL	PROCEDURE	VOICE
T–0:19:00		C2	On computer keyboard, enter: OPS 1 0 1 PRO. (You have just entered the OPS 101 program, which controls the Shuttle from T–20 minutes until lift-off, into the backup flight system).	
		C2	On computer keyboard, enter: SPEC 9 9 PRO.	
T–0:16:00	Main propulsion system (MPS) helium pressurization.	R2	MPS He ISOL (isolation) A, B switches (all six)—OPEN.	
			PNEUMATICS He ISOLATION switch —OP.	
T–0:15:00	Abort check.	F6	Mission control will cycle the ABORT light bright, dim, then off, three times.	G: _Columbia_, this is Control. We will conduct the abort check, over. Y: Roger: looks good, out.
T–0:09:00	10-minute hold (Allows one last chance for catch-up before you begin last part of your countdown).			
T–0:09:00	Resume countdown. Go for launch.	C2	EVENT TIMER switch to START.	
		F7	Check EVENT TIME indicator—it should be counting down. (Automatic ground launch sequencer starts at this time.)	Y: Control, this is _Columbia_. Event timer started, over. G: Roger, _Columbia_, out.
T–0:08:00	Alternating Current sensor to monitor.	R1	AC BUS SNSR switches (all three)—OFF for 1 second, then MONITOR.	
*T–0:07:00	Crew-access arm retracts.			
T–0:06:00	Auxiliary power unit (APU) prestart.		Prepare the three APUs, which power the Orbiter's hydraulic system for operation:	G: _Columbia_, this is Control. Initiate APU prestart procedure, over. Y: Roger, out.

TIME	EVENT	PANEL	PROCEDURE	VOICE
T–0:06:00		R2	Check the following switches to make sure they are in the positions indicated: BOILER N$_2$ SPLY (all three)—ON BOILER CNTLR (all three)—ON BOILER CNTLR PWR/HTR (all three)—A APU FUEL TK VLV (all three) —CL APU FUEL PUMP/VLV COOL (both) —OFF.	
		R2	APU CNTLR PWR (APU controller power) switches (all three)—ON.	
		R2	Check the following switches: HYD CIRC PUMP (hydraulic circulation pump)—GPC (controlled by general-purpose computer) APU AUTO SHUT DOWN—ENA APU SPEED SEL—NORM APU CONTROL—OFF.	
		R2	HYD MAIN PUMP PRESS (all three) (main pump pressure) selector switches—LO.	
T–0:05:00	Start APUs.	R2	APU FUEL TK VLV (APU fuel tank valves)—OPEN. APU CONTROL 1—START/RUN. HYD MAIN PUMP PRESS 1—NORM.	Y: Control this is *Columbia*. Pre-start complete. Powering up APUs, over. G: Roger, over. Y: APUs look good, out.
		F8	Check the HYDRAULIC PRESSURE 1 indicator—it should be HI green. *Repeat procedures for APUs #2 and #3.*	
		R2	HYD CIRC PUMP switches (all three)—OFF.	
		F7	Check HYD PRESS light—it should be off.	

TIME	EVENT	PANEL	PROCEDURE	VOICE	1.23
*T–0:04:30	Orbiter switches to internal power.			G:	*Columbia*, this is Control. You are on internal power, over.
				Y:	Roger, out.
*T–0:03:45	Orbiter aero surfaces are moved to condition the hydraulic system.			G:	*Columbia*, this is Control. Hydraulic check complete, over.
				Y:	Roger, out.
*T–0:03:00	Orbiter main engines gimbal (swivel) to their launch positions.			G:	*Columbia*, this is Control. Main engine gimbal complete, over.
				Y:	Roger, out.
*T–0:02:55	External Tank oxygen (O_2) vents close; liquid-oxygen tank begins pressurizing.			G:	*Columbia*, this is Control. O-two vents closed, looks good, over.
				Y:	Roger, out.
T–0:02:00	Configure for lift-off.	05, 09	LEFT (RIGHT) AUDIO MASTER VOL (audio volume control)—adjust to a comfortable level.		
		R2	APU AUTO SHUT DOWN (APU automatic shutdown) INHIBIT.	Y:	Control this is *Columbia*. APU to inhibit, over.
				G:	Roger, we copy *Columbia*, out.
*T–0:01:57	External Tank hydrogen (H_2) vents close; liquid-hydrogen tank pressure builds up for flight.			G:	*Columbia*, this is Control. H-two tank pressurization OK. You are go for launch, over.
				Y:	Roger, go for launch, out.
*T–0:00:25	Solid Rocket Booster APUs start. Management of countdown switches over to onboard computers.			G:	*Columbia*, this is Control. APU start is go. You are on your on-board computer, over.
				Y:	Roger, out.

TIME	EVENT	PANEL	PROCEDURE	VOICE 1.24
*T–0:0:03.8	Computers command Space Shuttle Main Engines Engines (SSMEs) to start.			
*T–0:0:03.46	First SSME ignites.			
*T–0:0:03.34	Second SSME ignites.			
*T–0:0:03.22	Third SSME ignites.			
T–0:0:00	Check main engine pressure.	F7	Check the MPS Press Pc gauge to make sure it reads higher than 90%.	G: Five—four—we have main engine start—two—one—zero—SRB ignition—lift off!
	Check main engine status.	F7	MN ENGINE STATUS lights (all three)—should be green.	
*T–0:0:00	2.64-second timer for Solid Rocket Booster (SRB) ignition starts.			
*T + 0:0:02.64	Solid boosters ignite.			
*T + 0:0:03	Lift-off.			
*T + 0:0:06.5	Launch tower cleared.			G: The tower has been cleared. All engines look good.
*T + 0:0:11	120° roll into "heads down" position starts.			G: Instituting roll maneuver.
*T + 0:0:30	Roll maneuver complete.			G: Roll maneuver complete. *Columbia,* you look good.
T + 0:0:30	Attitude direction indicator (ADI) to give local vertical/local horizontal (LVLH).	F6, F8	ADI ATT switches—LVLH positions	
*T + 0:0:44	(Time is approximate, depending on when you reach the speed of sound, Mach 1.) SSMEs throttle down from 100% thrust to 65%.			Y: Control, this is *Columbia.* Main engines at sixty-five percent, over. G: Roger, out.

TIME	EVENT	PANEL	PROCEDURE	VOICE 1.25
*T + 0:01:06	(Time depends on when you reach maximum dynamic pressure, "max q.") SSMEs throttle back up to 100%.			Y: Control, this is *Columbia*. Max Q, over. G: Roger, *Columbia*, out.
*T + 0:02:00	Solid booster burnout.			
T + 0:02:00	"Pc<50" signal.	CRT	The CRT displays report that the solid boosters have burned out by indicating their combustion chamber pressures (Pc) is less than 50 psi (345 kilopascals).	Y: Control, this is *Columbia*. We have SRB burnout; ready for SRB sep, over. G: Roger, out.
*T + 0:02:07	Solid Rocket Booster separation.			Y: Control, this is *Columbia*. We have SRB sep, over. G: Roger, we can see that, out.
*T + 0:04:20	Negative Return call from mission Control; a return-to-launch-site abort is no longer possible.			G: *Columbia*, this is Control. You are negative return. Do you copy? Over. Y: Roger, Mission Control, Negative return, out.
*T + 0:06:30	Shuttle begins a long shallow dive to prepare for External Tank separation.			
T + 0:07:00	Report to Mission Control that you can reach orbit even if two Main Engines fail.			Y: Control, this is *Columbia*. We are single engine press to MECO, over. G: Roger, *Columbia*, out.
*T + 0:07:40	Main Engines throttle down to keep acceleration less than 3 g.			G: *Columbia*, this is Control. Main Engine throttle down, over. Y: Roger, out.

|------|-------|-------|-----------|-------|------|
*T + 0:08:28	Engines throttle down to 65% of thrust.			G: *Columbia*, this is Control. Go for Main Engine cut-off, over. Y: Roger, Main Engine cut-off on schedule, out.	
*T + 0:08:38	Main Engine cut-off (MECO); the three main engines shutdown.				
T + 0:08:38	MECO indicated on instrument panel.	F7	MAIN ENGINE STATUS indicators (all three)—should be red, indicating the engines have stopped.		
T + 0:08:54	External Tank separation; check that MAIN ENGINE STATUS lights are off.			G: *Columbia*, this is Control. Go for ET separation. Y: Roger, we have External Tank sep.	
T + 0:09:00	−Z translation complete; prepare for orbital maneuvering system burn #1 (OMS-1).			Y: (continued) Beginning minus Z translation, out.	
T + 0:10:39	Orbital Maneuvering System burn #1.	F6, F8	ADI ATTITUDE—INRTL (inertial).	G: *Columbia*, this is Control. You are go for OMS-one burn, over. Y: Roger, OMS-one, out.	
		C3	DAP (digital autopilot)—AUTO. On computer keyboard, enter: ITEM 2 7 EXEC.		

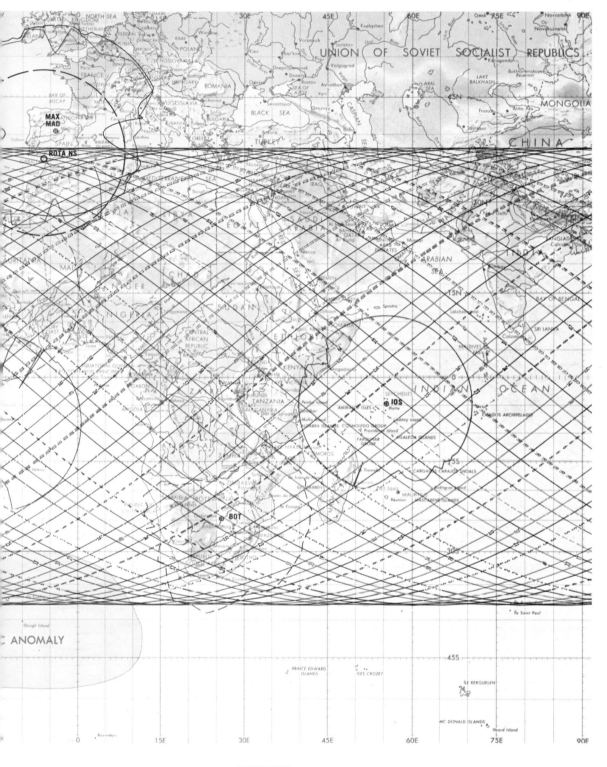

GROUNDTRACK DATA

FRONT

Orbits 1 through 16 are shown
by numerals corresponding to
the orbit being depicted ....

Orbits 17 through 32 ....

Orbits 33 through 48 ....

Orbits 49 through 64 ....

BACK

Orbits 65 through 80 ....

Orbits 81 through 96 ....

Orbits 97 through 112 ......

Orbits 113 through 116 ......

SPACE SHUTTLE
MISSION CHART
STS-3
FRONT – ORBITS 1 TO 64

EDITION - 1   JANUARY 1982

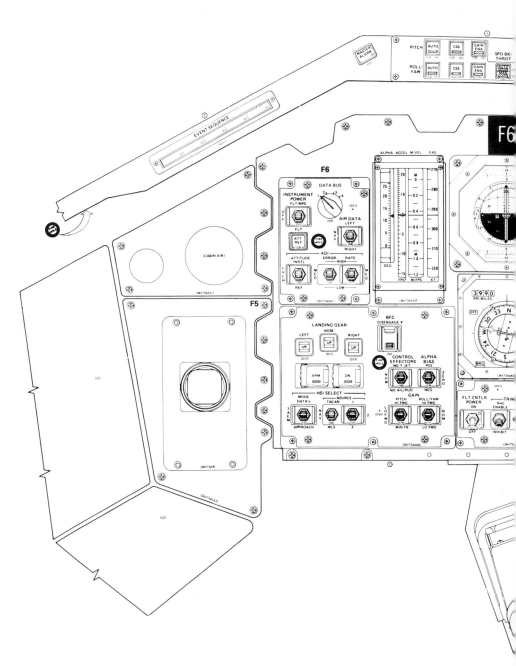

CONTROL STICK

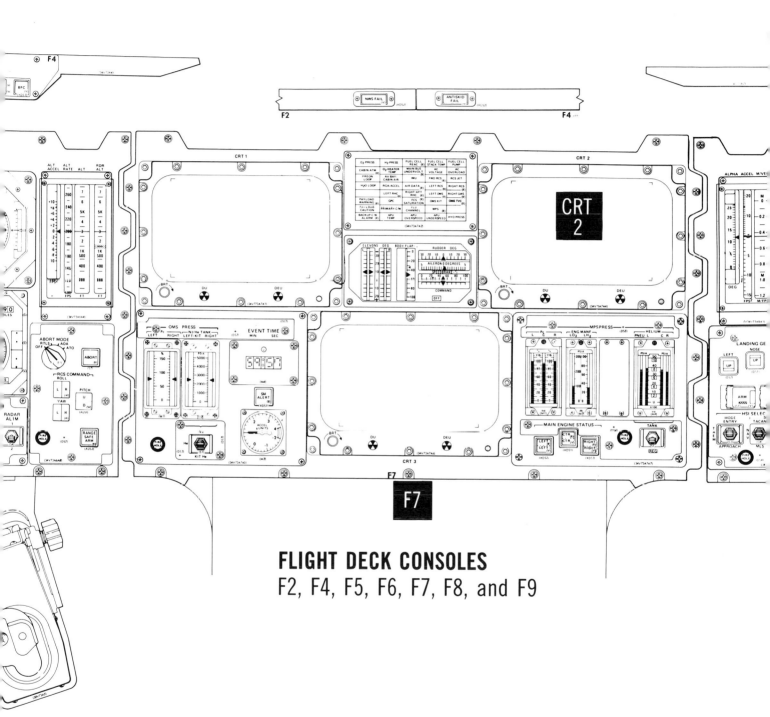

FLIGHT DECK CONSOLES
F2, F4, F5, F6, F7, F8, and F9

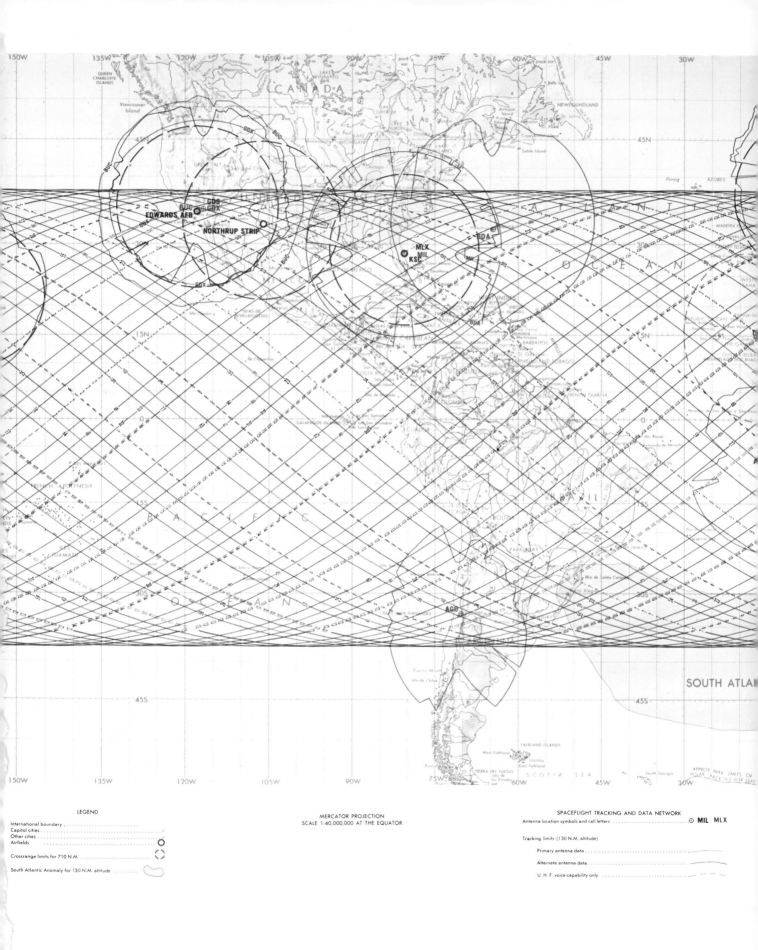

LEGEND

International boundary . . . .
Capitol cities . . . .
Other cities . . . .
Airfields . . . .

Crossrange limits for 710 N.M. . . . .

South Atlantic Anomaly for 130 N.M. altitude . . . .

MERCATOR PROJECTION
SCALE 1:40,000,000 AT THE EQUATOR

SPACEFLIGHT TRACKING AND DATA NETWORK

Antenna location symbols and call letters . . . . ⊕ MIL MLX

Tracking limits (130 N.M. altitude)

Primary antenna data . . . .

Alternate antenna data . . . .

U. H. F. voice capability only . . . .

GROUND TRACK OF THE FIRST HALF OF THE 7-DAY STS-3 FLIGHT.

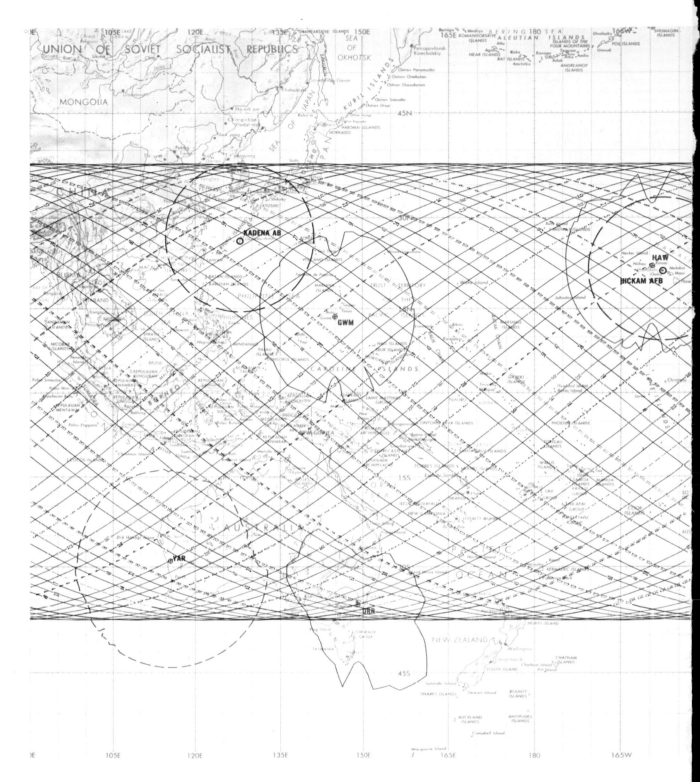

PREPARED AND PUBLISHED BY THE DEFENSE MAPPING AGENCY
AEROSPACE CENTER, ST. LOUIS AFS, MISSOURI 63118 FOR THE NATIONAL
AERONAUTICS AND SPACE ADMINISTRATION

NOTE: The representation of international
boundaries on this chart is not necessarily
authoritative.

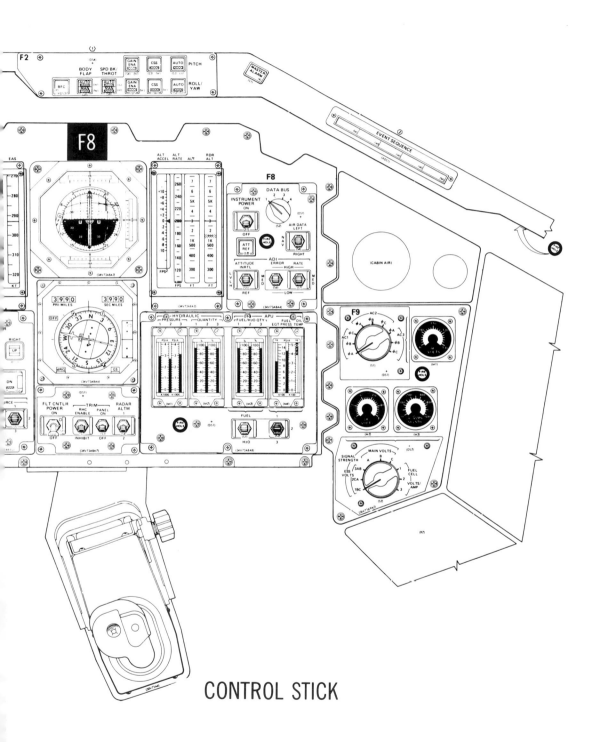

CONTROL STICK

| --- | --- | --- | --- | --- | --- |
*T + 0:12:24	OMS-1 cut-off.			Y: Control, this is *Columbia*. We have OMS cut-off, over. G: Roger, we copy. Please advise when ET umbilical doors are closed, over.	
T + 0:12:30	Post-OMS-1 activities. Auxiliary power unit shutdown.	R2	APU AUTO SHUT DN——ENA. BOILER CNTLR——OFF. APU CONTROL——OFF.		
	Change computer program.	C2	On computer keyboard, enter: OPS 1 0 5 PRO.		
	Close and latch External Tank umbilical doors.	R2	ET UMBILICAL DOOR MODE——GPC/MAN (computer or manual) CENTERLINE LATCH——STO. L and R DOOR——OP. L and R DR LATCH switches——LAT.	Y: Umbilical doors closed, over. G: Roger, out.	
T + 0:45:58	Orbital maneuvering system burn #2 (OMS-2).	08	OMS FNG (OMS engine arming switch)——ARM/PRESS.	G: *Columbia*, this is Control. Coming up on OMS-two, over. Y: Roger, OMS-two.	
T + 0:46:34	OMS-2 cut-off.			Y: OMS-two cut-off. We have achieved orbit, over.	
	Change computer program.	C2	On computer keyboard, enter: OPS 1 0 6 PRO.	G: Roger, *Columbia*, out.	

CONGRATULATIONS, YOU ARE IN ORBIT.

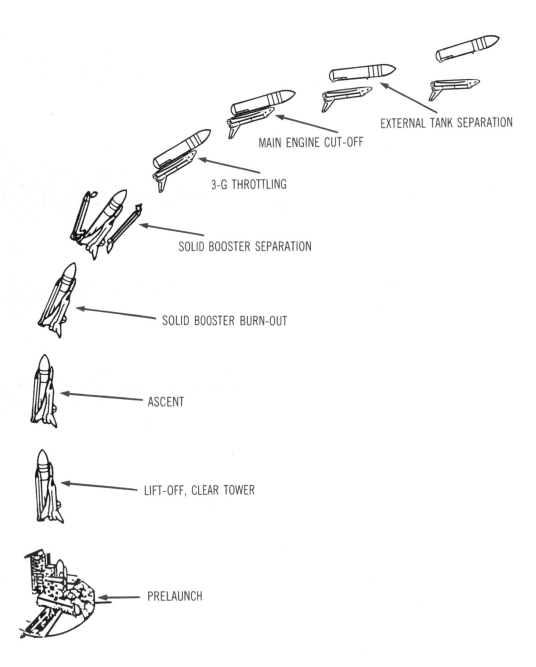

EXTERNAL TANK SEPARATION

MAIN ENGINE CUT-OFF

3-G THROTTLING

SOLID BOOSTER SEPARATION

SOLID BOOSTER BURN-OUT

ASCENT

LIFT-OFF, CLEAR TOWER

PRELAUNCH

For major maneuvers such as the transfer from one orbit to another, the 12,000-pound-thrust (53,000-newton) **orbital maneuvering system** (OMS) is used. Small maneuvers use clusters of rocket engines in the Orbiter's nose and tail. Called the **reaction control system** (RCS), they can move the Orbiter about its three major axes (called **rotation**) or in a straight line (**translation**).

The forward RCS engines are on the top and sides of the Orbiter, between the nose and cockpit. Both aft OMS pods have rearward extensions containing rows of RCS motors. Two types of engines are used: the primary engines with 870 pounds of thrust (3,800 new-

tons) each, and smaller 24-pound-thrust (105-newton) vernier engines for precise adjustments and corrections. Designed to last as long as the Orbiter, each RCS engine can be fired fifty thousand times over one hundred flights.

The forward and aft RCS engines point in different directions. Using the correct combination permits you to move the Orbiter in any direction. The system which controls the engine is called the **digital auto pilot**, or **DAP** for short. There are two sets of controls for the DAP: one in your center console, panel C3, and one on aft crew station panel A6.

DIGITAL AUTOPILOT CONTROLS

You can select either automatic (AUTO) or manual (MAN) modes for the DAP. A light on the switch tells you which is on. Below these switches, two groups of nine switches each control translation and rotation manual operations. The pair of switches to the right of the AUTO and MAN switches allow you to select either the primary or vernier RCS engines.

To rotate the Orbiter, first select MAN mode and NORM RCS jets. For a **roll** maneuver, push the DISC RATE button under the ROLL heading. (This selects the rotation rate indicated by your onboard computer.) Next, grasp the rotational hand controller in front of you and move it left or right, depending on the direction you want to roll. For **pitch** and **yaw**, after selecting the axis and rate on the DAP controls, move the hand controller forward or backward (pitch up or down) or twist it left or right (yaw left or right).

Translation maneuvers are made in much the same way. Select the DAP control mode, RCS jets, and maneuver rate along the X, Y, or Z axis. There are only two translation hand controllers in the Orbiter. One is on the left side of the commander's forward panel, the other is on the aft crew station. Pushing or pulling the translation controller moves you forward or backward (**X-axis**). Left or right motion moves you sideways (**Y-axis**), and moving the handle up or down moves you up or down along the **Z-axis**.

Maneuvering in Space

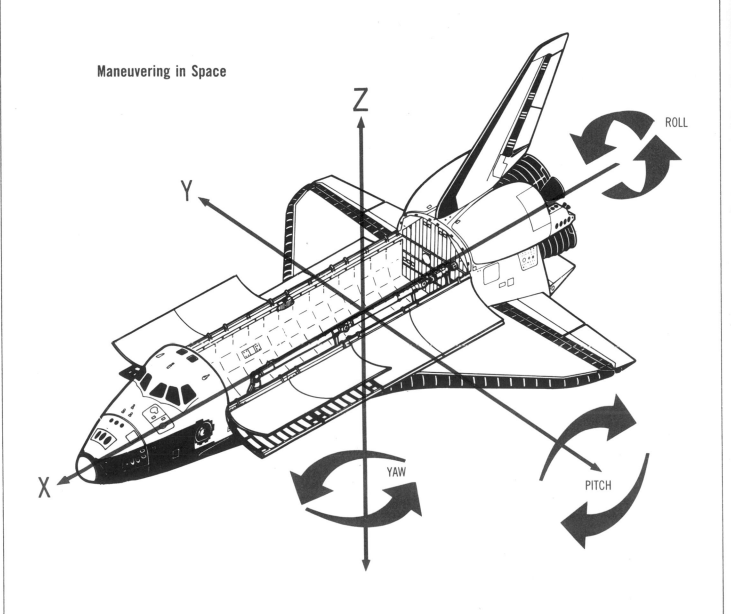

The center of the Orbiter's fuselage comprises the **cargo bay**. It's frequently called the **payload bay**, but cargo bay is the preferred name. That's because the more general term "cargo" encompasses not only the useful **payload** carried but also payload **support hardware**. Support hardware includes launch cradles, environmental-control equipment, and prerelease-test equipment. Such equipment doesn't contribute to the work performed by the payload, but it is necessary, has weight, and must be considered when computing cargo weight. The cargo bay is 15 feet (4.5 meters) in diameter and 60 feet (18 meters) long, large enough to hold a tour bus with room to spare.

As soon as you're in orbit and have checked the Orbiter, you must open the two curved **cargo-bay doors**. **Radiators** to rid the craft of excess heat are built into the inner surface of the doors. If the doors remained closed, heat would build up inside the Orbiter and you'd have to land after only a few hours. You'll have to close the doors again just before retro-fire. When they're closed, thirty-two latches secure the doors.

The controls for the cargo-bay doors are located on panel R13 in the aft area of the flight deck.

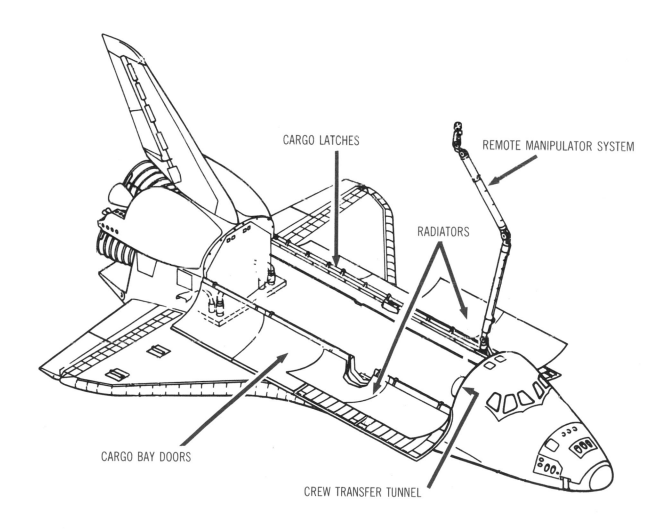

CARGO LATCHES

REMOTE MANIPULATOR SYSTEM

RADIATORS

CARGO BAY DOORS

CREW TRANSFER TUNNEL

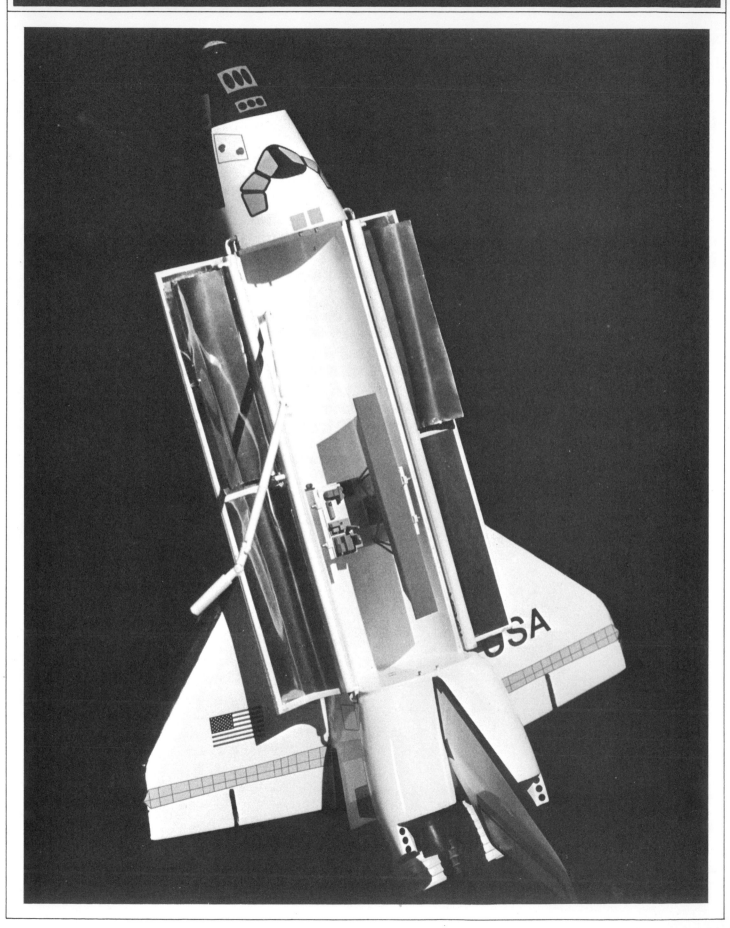

This section describes how you accomplish such everyday activities as eating and sleeping in space. Not only is there no air, food, or water in space, requiring that all this be carried with you, but the environment is extremely inhospitable. Temperatures can vary from 120°C (250°F) in the sun to -100°C (-150°F) in the shade, so one side of you bakes while the other freezes. There are cosmic rays, solar flares, and other harmful forms of radiation. You can add to the list such other hazards as micrometeroids, physical deterioration from weightlessness, and psychological conflicts caused by being confined in a small space.

However, these problems were considered when the Orbiter was designed, so you'll find you can live aboard the Shuttle in relative safety and comfort.

The Space Shuttle's **life-support system** automatically supplies air and water.

An atmosphere like the earth's is maintained in the crew compartment. Atmospheric pressure is 14.7 psi (pounds per square inch) (100 kilopascals), the same as standard sea-level conditions. The atmosphere consists of 79% nitrogen and 21% oxygen, again very close to what you left on earth. In an emergency, cabin pressure is reduced to 8.0 psi (55 kilopascals).

The **cabin-pressurization system** has two oxygen supplies, two nitrogen supplies, and an emergency oxygen system. You use about 1.8 pounds (.8 kilograms) of oxygen per day. For a five-person crew, a consumption of up to 7.7 pounds (3.5 kilograms) of nitrogen and 9 pounds (4 kilograms) of oxygen per day is normal. However, having enough oxygen isn't the only consideration. Your body converts oxygen to carbon dioxide. Carbon dioxide is a waste product and must not be allowed to build up in the cabin.

Therefore, canisters filled with lithium hydroxide and activated charcoal remove carbon dioxide and odors from the cabin's air. Fans circulate the air around the cabin and through the canisters, which are located beneath the mid-deck floor. Carbon dioxide reacts with lithium hydroxide to produce lithium carbonate and water vapor.

Two canisters are used at a time. With four people aboard, an individual canister lasts 24 hours. They are changed alternately; that is, one is changed every 12 hours. As with other chores, responsibility for changing the carbon dioxide absorbers is assigned to a different crew member each day.

When it is your turn, you will find it takes only a few minutes. Access panels for the life-support system and the fresh-canister stowage locker are next to each other on the mid-deck floor. Both are clearly marked. After opening the stowage locker, remove a fresh canister. Next, open the life-support system access panel, remove the old unit, and replace it with the new one. Close the panel and dispose of the expended canister with the dry trash.

The cabin's atmosphere is controlled from panels L1 and L2, the commander's left-hand consoles. Panel L2 contains the atmospheric pressure controls. There are switches to select which source of oxygen you want to use. These are marked O_2 SYS 1 (or 2) SUPPLY for the primary and secondary systems, and O_2 EMER for the emergency supply. The supply switch for the system you are using should be in the OPEN position, the others should be at CLOSE. Switches for the nitrogen supplies, marked N_2 SYS 1 (or 2) are on the same panel.

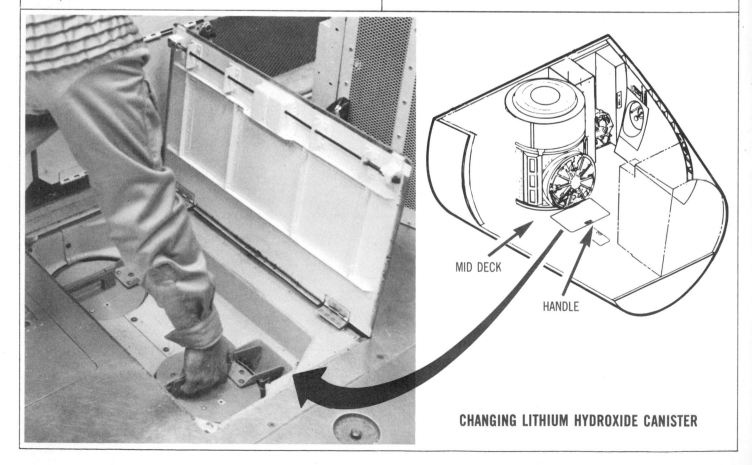

MID DECK

HANDLE

CHANGING LITHIUM HYDROXIDE CANISTER

Your oxygen supply is carried in the form of a super-cold (**cryogenic**) liquid. Liquid oxygen boils at −118°C (−180°F). Each liquid-oxygen tank has heaters which vaporize some of the super-cold liquid. This way, pressure inside the tanks is kept between 835 and 852 psi (5,757 and 5,875 kilopascals). When you open an oxygen-supply valve, cold oxygen gas from the tank flows through a heat exchanger, where it is warmed before passing through the system's regulator. By the time the oxygen reaches the O_2/N_2 CNTRL VLV (**oxygen/nitrogen control valve**), its pressure is 100 psi (690 kilopascals).

The nitrogen system has four tanks, which contain nitrogen gas at a pressure of 3,300 psi (22,700 kilopascals). There are two tanks for each system (primary and secondary). Nitrogen flows from its tank through a regulator where its pressure is reduced to 200 psi (1,380 kilopascals). The gas then enters the oxygen/nitrogen control valve.

Sensors monitor the oxygen content of the cabin atmosphere. When the PPO_2 SNSR/VLV (**partial pressure of oxygen sensor/valve**) switch on panel L2 is on NORM and the O_2/N_2 CNTRL VLV is on AUTO, the gas-control valves and oxygen sensors work together. If the sensors indicate that oxygen is needed in the cabin, the nitrogen valve closes and oxygen flows through the cabin regulator. When there is enough oxygen, the nitrogen valve opens and closes the oxygen valve. This way, the partial pressure of oxygen in the cabin is maintained at 3.45 psi (24 kilopascals).

If you get too warm or too cold, you can adjust the cabin temperature using the CABIN TEMP switch on panel L1. Below this switch are controls for the fans that circulate air through the lithium hydroxide canisters.

Electrical power for the Orbiter is generated by **fuel cells**. Oxygen and hydrogen are chemically combined in the cells to produce electricity. There is an important by-product from this reaction—water. About 7 pounds (3 kilograms) of water is created every hour. This water is fed to a pair of tanks, each of which can hold 165 pounds (75 kilograms) of water. One tank is for the spacecraft thermal-control system, the other provides water to the galley. When the tanks are full, water from the fuel cells is dumped overboard where it instantly freezes, then becomes a gas. Without pressure it sublimates cleanly and harmlessly.

NITROGEN SUPPLY SYSTEM SELECTION

TURNING ON CABIN FAN

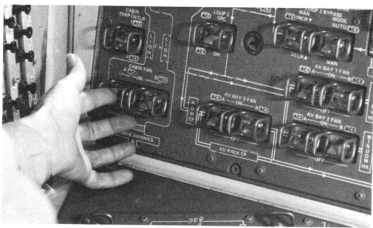

ADJUSTING CABIN TEMPERATURE

QUICK CHECKLIST
FOR ATMOSPHERIC CONTROL PANELS

PANEL L2

O_2 SYS 1 SUPPLY	OPEN
O_2 EMER	CLOSE
O_2 SYS 2 SUPPLY	CLOSE
O_2 X OVER SYS 1	CLOSE
O_2 X OVER SYS 2	CLOSE
N_2 SYS 1	
SUPPLY	OPEN
REG INLET	OPEN
N_2 SYS 2	
SUPPLY	CLOSE
REG INLET	CLOSE
PpO_2 SNSR/VLV	NORM
O_2/N_2 CNTRL VLV SYS 1	OPEN
O_2/N_2 CNTRL VLV SYS 2	CLOSE

PANEL L1

CABIN FAN	
A	ON
B	OFF
CABIN TEMP	Adjust for comfort
HUMIDITY SEP	
A	ON
B	OFF

TEMPERATURE CONTROL

WATER CIRCULATES THROUGH A SERIES OF PIPES AND HEAT EXCHANGERS TO HELP REMOVE UNWANTED HEAT FROM THE CABIN. THE WATER IS PUMPED INTO THE MID-FUSELAGE WHERE THE HEAT IS PASSED ON TO THE FREON-21 RADIATORS IN THE CARGO BAY DOORS.

CABIN HEAT EXCHANGER

WATER PUMPS

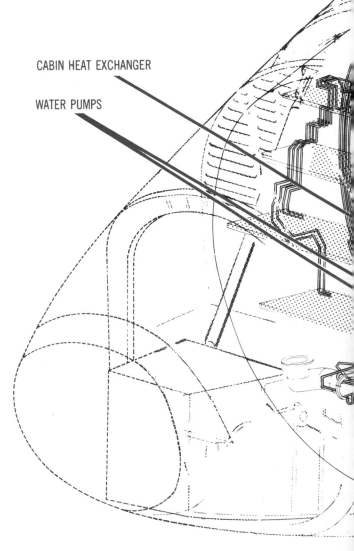

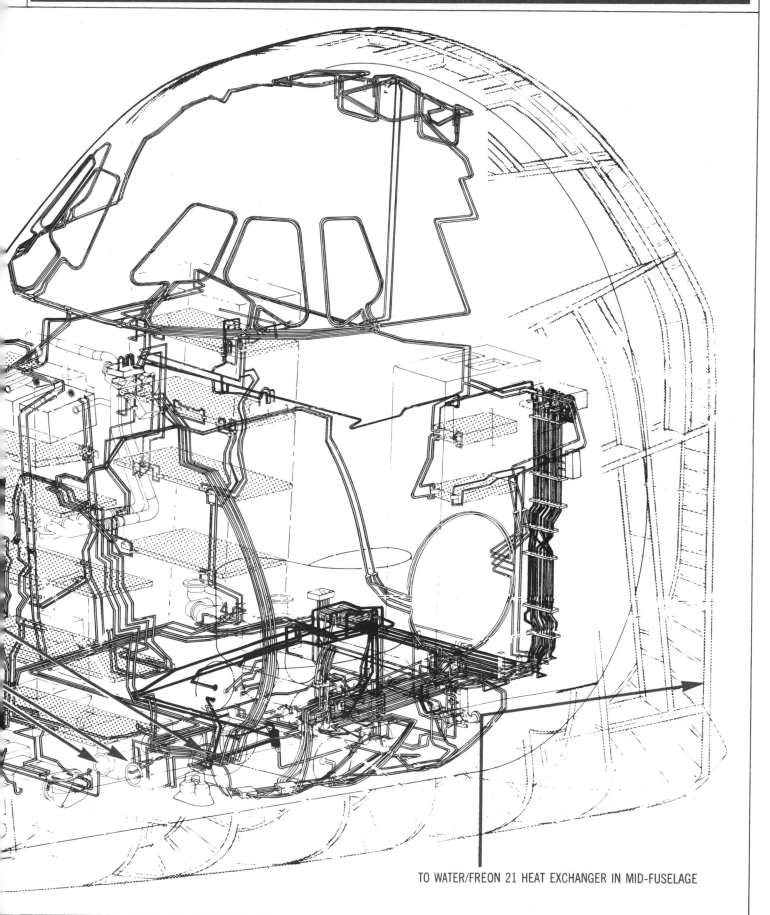

TO WATER/FREON 21 HEAT EXCHANGER IN MID-FUSELAGE

In the first 10 minutes of flight you experience two stresses which, although they are opposites, are related. The first is the sensation of increased weight during ascent; the other is your apparent lack of weight in orbit.

During launch, the Shuttle gains speed or accelerates. This acceleration pushes you back into your seat, making you feel heavier. Launch accelerations are expressed in terms of gravities, or g-forces. One g is your normal body weight on earth. Accelerating at 2 g makes you feel as though your weight has doubled. The Shuttle reaches a peak acceleration of 3 g. At lift-off, you are in a sitting position on your back, so g-forces are exerted against your chest, along the X-axis of the Shuttle. Three g is less than half the acceleration experienced by astronauts during the Apollo flights of a decade ago. (In fact, 3 g is less than you experience on some roller coasters.) Launch accelerations are kept low by reducing the thrust of the Shuttle's engines. The relatively low acceleration means that any normal, healthy individual can fly on the Shuttle.

To achieve orbit, the Shuttle must reach a speed of 17,500 mph (28,000 kph). Left unchecked, your Orbiter would continue in a straight line off into space. This property, shared by all objects, is called **inertia**. However, earth's gravity pulls the craft down. At this speed, the Orbiter's inertia (tendency to fly off into space) and the pull of earth's gravity are equal. Therefore, the Shuttle "falls" around the earth and will continue to do so until you fire a maneuvering rocket and change the Orbiter's velocity.

In orbit, you experience what has been called **weightlessness**. Actually, objects in orbit still have **some** weight. This is important to designers of some experiments, but the amount is so slight you won't be able to feel it. However, to be precise, the term **microgravity** has been adopted to describe your lack of weight. Microgravity affects your body in several ways.

One of the most noticeable is that the fluids in your body redistribute themselves. This means that the fluid content in your upper body increases and causes your face to puff out. Also, you may experience some fullness or stuffiness in your sinuses.

Before launch, you were issued what looked like a Band-Aid and were told to place it behind your right ear. It contains a medication for motion sickness, which is absorbed through the skin. This time-release medicine prevents discomfort, or **space sickness**. In the past, some astronauts had a hard time getting used to microgravity and were nauseated for their first few days in orbit. Since your flight will only last a few days, we want you to be in peak physical shape. The medication should help you adjust to your new environment.

Another, more subtle, effect you'll experience is the way your posture alters in microgravity. If you relax and allow yourself to just float, you'll naturally assume a **neutral body position**. In this posture, your arms float away from and just in front of your body. You're bent slightly at the waist, your knees are flexed, and your toes point a bit. The toe pointing makes it hard for you to stand erect with your feet flat on the floor. To correct this, special attachments for your boots have been developed. When worn, the shoe attachments elevate your heel so you can comfortably stand straight. Further, each has two suction cups to anchor you to the floor.

MICROGRAVITY SHOE ATTACHMENTS ON VARIABLE-HEIGHT WORK PLATFORM. THESE ATTACHMENTS HAVE SUCTION CUPS TO ANCHOR YOU TO PLATFORM OR FLOOR.

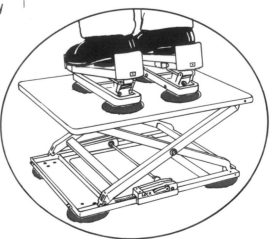

NEUTRAL BODY POSITION

The effects of microgravity on your heart are similar to those of prolonged bed rest. To prevent your heart from weakening, you have to exercise each day. A treadmill made from a piece of Teflon attached to an aluminum base is available for that purpose. Rubber bungee cords attached to a belt and shoulder harness hold you down. You can adjust the tension of the bungees. When it's your time to exercise, unstow the treadmill and attach it to a convenient floor or wall (remember, you're weightless). Put on the harness, adjust the tension, and run in place. The tighter the tension, the harder the exercise becomes. Fifteen minutes a day is enough for flights lasting less than a week. On longer missions, increase your exercise period to 30 minutes.

The Orbiter loses speed quickly as it plunges into the atmosphere during entry. This rapid change in speed, or **deceleration** (since you're slowing down), creates a stress of 1.5 g on your body. However, during entry the g-force is exerted in the −Z axis of the Shuttle, or from your head to your toes. When experiencing g-forces in this direction, blood flows away from your brain and pools in the lower half of your body. This can cause grayout, or even blackout, when you completely pass out. To prevent this, you must wear a special garment called an anti-g suit. The anti-g suit resembles a pair of trousers. As g-forces build up, bladders inside the trousers inflate and place pressure on your lower body, literally squeezing blood into your upper body. Blackouts during this most critical part of your flight are thus prevented.

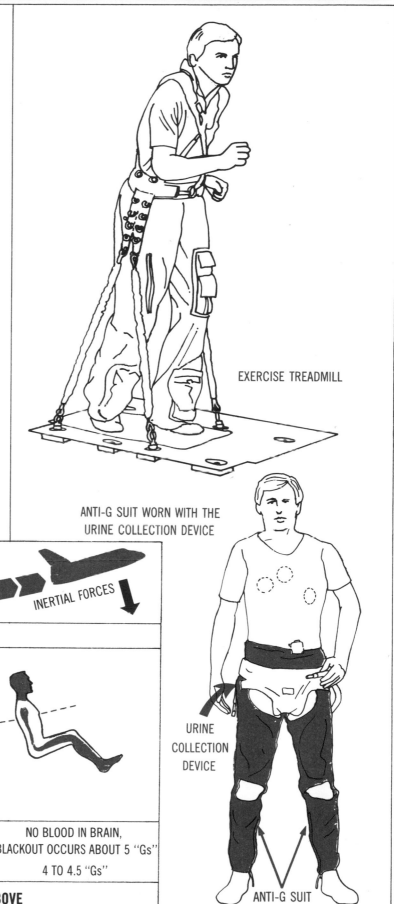

EXERCISE TREADMILL

ANTI-G SUIT WORN WITH THE
URINE COLLECTION DEVICE

URINE
COLLECTION
DEVICE

ANTI-G SUIT

POSITIVE ACCELERATION (SUSTAINED)

INERTIAL FORCES

EFFECT ON BODY

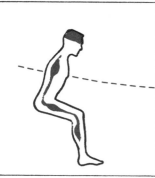

AS ACCELERATION STARTS, BLOOD BEGINS TO POOL
1 TO 2.5 "Gs"

POOLING INCREASES, VISION BEGINS TO FADE
(GRAYOUT) 2.5 TO 4 "Gs"

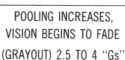

NO BLOOD IN BRAIN, BLACKOUT OCCURS ABOUT 5 "Gs"
4 TO 4.5 "Gs"

THE ANTI-G SUIT PREVENTS THE ABOVE

The food system aboard the Space Shuttle has been designed to provide meals that are both nutritious and appetizing. During early space flights, meals were restricted to items that could be puréed and placed in metal squeeze tubes or compressed into bite-size tablets. Needless to say, reaction to early space food was less than enthusiastic. The Gemini program saw the introduction of **freeze-dried** foods that could be rehydrated. Moon-bound Apollo astronauts also had canned food items and fresh bread. Another innovation introduced on Apollo was **spoon-bowl** packaging. Some dehydrated foods were packaged in bags with zip-lock seals. Although freeze-drying was an improvement over the squeeze tubes, food still had to be squeezed through a slit in one end of the bag. With the spoon-bowl, after water is added through a valve in the bottom of the package, the zipper is opened and the food eaten with a spoon. Skylab had a freezer, so astronauts aboard the orbiting workshop enjoyed filet mignon, lobster Newburg, and vanilla ice cream.

Individual food items for Skylab were packed in aluminum cans with pull-off lids. Another new feature was the use of a table and individual food trays with eight cavities for the food cans. Three cavities had heaters beneath them, which warmed the food to 66° C (150°F). The surface of each tray was magnetized to hold the utensils—knife, fork, spoon, and scissors—in place.

Space flight cuisine has come of age on the Shuttle. The Orbiter has a combination **galley** and **personal-hygiene station** in the mid-deck, near the entry hatch. The galley has hot and cold water dispensers, serving trays, a pantry, and a convection oven for warming food. Unlike Skylab, the Orbiter has no freezers or refrigerators.

Your food is preserved in several ways. Of course some items are dehydrated or freeze-dried. Just add the amount of water indicated on the package to return the product to its natural form. Such foods as sliced beef, ham, and turkey with gravy are packed ready-to-eat in foil pouches. Those meats and other canned goods which don't have to be rehydrated are **thermostabilized**. This means that when they were packaged, they were heated enough to kill bacteria and prevent spoilage. Breads and some meats are **irradiated**, that is, treated with radiation, to kill bacteria. Finally, you'll find such snacks as nuts, granola bars, and even Life Savers candies in their natural form.

All the assorted food packages—cans, individual plastic serving containers, and foil pouches—fit into your food tray. Foods that need to be heated are packed in either plastic containers or foil pouches. Both fit in the oven.

The menu is a 6-day standard menu, and everybody has the same meal. However, if you don't want a particular item, you can exchange it for something from the pantry. Your diet has been constructed to give you about 3,000 calories per day.

Crew members take turns preparing meals. When it's your turn, you'll find it takes only about 20 minutes to prepare a meal for a crew of seven. Before launch, individual servings were assembled into meals, over-wrapped, and packed in pouches. Pouches are marked by day and meal. After you find the correct pouch, unwrap and remove the servings. Add water to the items that need to be reconstituted and place the articles to be warmed in the oven. While these are heating, add water to the beverage containers. When everything is ready, assemble the cans, containers, and foil packs into meals on the individual food trays.

The food trays have magnets on them so you can attach yours to the work/dining table in the mid-deck or even the cabin wall. Open the food packages with your scissors—and **bon appétit**!

After the meal, stow the empty food containers in the wet-trash compartment and clean your utensils and food trays with germicidal wet wipes so they can be reused.

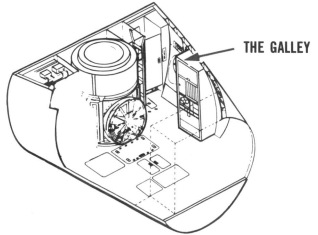

THE GALLEY

SPACE SHUTTLE FOOD[a] AND BEVERAGE LIST

APPLESAUCE (T)	CHICKEN AND NOODLES (R)	PEACHES, DRIED (IM)
APRICOTS, DRIED (IM)	CHICKEN AND RICE (R)	PEACHES, (T)
ASPARAGUS (R)	CHILI MAC W/BEEF (R)	PEANUT BUTTER
BANANAS (FD)	COOKIES, PECAN (NF)	PEARS (FD)
BEEF ALMONDINE (R)	COOKIES, SHORTBREAD (NF)	PEARS (T)
BEEF, CORNED (I) (T)	CRACKERS, GRAHAM (NF)	PEAS W/BUTTER SAUCE (R)
BEEF AND GRAVY (T)	EGGS, SCRAMBLED (R)	PINEAPPLE, CRUSHED (T)
BEEF, GROUND W/PICKLE SAUCE (T)	FOOD BAR, ALMOND CRUNCH (NF)	PUDDING, BUTTERSCOTCH (T)
BEEF JERKY (IM)	FOOD BAR, CHOCOLATE CHIP (NF)	PUDDING, CHOCOLATE (R) (T)
BEEF PATTY (R)	FOOD BAR, GRANOLA (NF)	PUDDING, LEMON (T)
BEEF, SLICES W/BARBEQUE SAUCE (T)	FOOD BAR, GRANOLA/RAISIN (NF)	PUDDING, VANILLA (R) (T)
BEEF STEAK (I) (T)	FOOD BAR, PEANUT BUTTER/GRANOLA (NF)	RICE PILAF (R)
BEEF STROGANOFF W/NOODLES (R)	FRANKFURTERS (VIENNA SAUSAGE) (T)	SALMON (T)
BREAD, SEEDLESS RYE (I) (NF)	FRUITCAKE	SAUSAGE PATTY (R)
BROCCOLI AU GRATIN (R)	FRUIT COCKTAIL (T)	SHRIMP CREOLE (R)
BREAKFAST ROLL (I) (NF)	GREEN BEANS, FRENCH W/MUSHROOMS (R)	SHRIMP COCKTAIL (R)
CANDY, LIFE SAVERS, ASSORTED FLAVORS (NF)	GREEN BEANS AND BROCCOLI (R)	SOUP, CREAM OF MUSHROOM (R)
CAULIFLOWER W/CHEESE (R)	HAM (I) (T)	SPAGHETTI W/MEATLESS SAUCE (R)
CEREAL, BRAN FLAKES (R)	JAM/JELLY (T)	STRAWBERRIES (R)
CEREAL, CORNFLAKES (R)	MACARONI AND CHEESE (R)	TOMATOES, STEWED (T)
CEREAL, GRANOLA (R)	MEATBALLS W/BARBECUE SAUCE (T)	TUNA (T)
CEREAL, GRANOLA W/BLUEBERRIES (R)	NUTS, ALMONDS (NF)	TURKEY AND GRAVY (T)
CEREAL, GRANOLA W/RAISINS (R)	NUTS, CASHEWS (NF)	TURKEY, SMOKED/SLICED (I) (T)
CHEDDAR CHEESE SPREAD (T)	NUTS, PEANUTS (NF)	TURKEY TETRAZZINI (R)
CHICKEN A LA KING (T)	PEACH AMBROSIA (R)	VEGETABLES, MIXED ITALIAN (R)

APPLE DRINK	INSTANT BREAKFAST, VANILLA	GRAPEFRUIT DRINK
COCOA	LEMONADE	INSTANT BREAKFAST, CHOCOLATE
COFFEE, BLACK	ORANGE DRINK	INSTANT BREAKFAST, STRAWBERRY
COFFEE W/CREAM	ORANGE-GRAPEFRUIT DRINK	TEA W/LEMON AND SUGAR
COFFEE W/CREAM AND SUGAR	ORANGE-PINEAPPLE DRINK	TEA W/SUGAR
COFFEE W/SUGAR	STRAWBERRY DRINK	TROPICAL PUNCH
GRAPE DRINK	TEA	

[a]ABBREVIATIONS IN PARENTHESES INDICATE TYPE OF FOOD: T = THERMOSTABILIZED, I = IRRADIATED, IM = INTERMEDIATE MOISTURE, FD = FREEZE DRIED, R = REHYDRATABLE, AND NF = NATURAL FORM.

A TYPICAL SPACE SHUTTLE MENU

DAY 1	DAY 2	DAY 3	DAY 4
PEACHES (T)	APPLESAUCE (T)	DRIED PEACHES (IM)	DRIED APRICOTS (IM)
BEEF PATTY (R)	BEEF JERKY (NF)	SAUSAGE (R)	BREAKFAST ROLL (I) (NF)
SCRAMBLED EGGS (R)	GRANOLA (R)	SCRAMBLED EGGS (R)	GRANOLA W/BLUEBERRIES (R)
BRAN FLAKES (R)	BREAKFAST ROLL (I) (NF)	CORNFLAKES (R)	VANILLA INSTANT BREAKFAST (B)
COCOA (B)	CHOCOLATE INSTANT BREAKFAST (B)	COCOA (B)	GRAPEFRUIT DRINK (B)
ORANGE DRINK (B)	ORANGE-GRAPEFRUIT DRINK (B)	ORANGE-PINEAPPLE DRINK (B)	
FRANKFURTERS (T)	CORNED BEEF (T) (I)	HAM (T) (I)	GROUND BEEF W/
TURKEY TETRAZZINI (R)	ASPARAGUS (R)	CHEESE SPREAD (T)	PICKLE SAUCE (T)
BREAD (2) (I) (NF)	BREAD (2) (I) (NF)	BREAD (2) (I) (NF)	NOODLES AND CHICKEN (R)
BANANAS (FD)	PEARS (T)	GREEN BEANS AND BROCCOLI (R)	STEWED TOMATOES (T)
ALMOND CRUNCH BAR (NF)	PEANUTS (NF)	CRUSHED PINEAPPLE (T)	PEARS (FD)
APPLE DRINK (2) (B)	LEMONADE (2) (B)	SHORTBREAD COOKIES (NF)	ALMONDS (NF)
		CASHEWS (NF)	STRAWBERRY DRINK (B)
		TEA W/LEMON AND SUGAR (2) (B)	
SHRIMP COCKTAIL (R)	BEEF W/BARBEQUE SAUCE (T)	CREAM OF MUSHROOM SOUP (R)	TUNA (T)
BEEF STEAK (T) (I)	CAULIFLOWER W/CHEESE (R)	SMOKED TURKEY (T) (I)	MACARONI AND CHEESE (R)
RICE PILAF (R)	GREEN BEANS W/MUSHROOMS (R)	MIXED ITALIAN VEGETABLES (R)	PEAS W/BUTTER SAUCE (R)
BROCCOLI AU GRATIN (R)	LEMON PUDDING (T)	VANILLA PUDDING (T) (R)	PEACH AMBROSIA (R)
FRUIT COCKTAIL (T)	PECAN COOKIES (NF)	STRAWBERRIES (R)	CHOCOLATE PUDDING (T) (R)
BUTTERSCOTCH PUDDING (T)	COCOA (B)	TROPICAL PUNCH	LEMONADE (B)
GRAPE DRINK (B)			

CONDIMENTS

PEPPER	BARBEQUE SAUCE	HOT PEPPER SAUCE	MUSTARD
SALT	CATSUP	MAYONNAISE	

[a]ABBREVIATIONS IN PARENTHESES INDICATE TYPE OF FOOD: T = THERMOSTABILIZED, I = IRRADIATED, IM = INTERMEDIATE MOISTURE, FD = FREEZE DRIED, R = REHYDRATABLE, NF = NATURAL FORM, AND B = BEVERAGE.

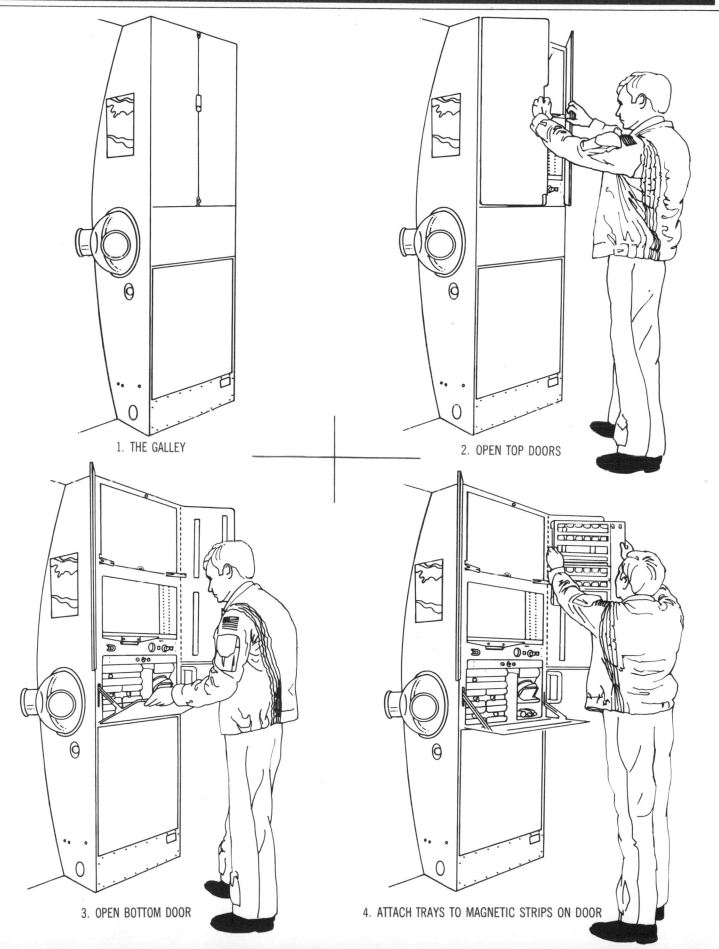

1. THE GALLEY

2. OPEN TOP DOORS

3. OPEN BOTTOM DOOR

4. ATTACH TRAYS TO MAGNETIC STRIPS ON DOOR

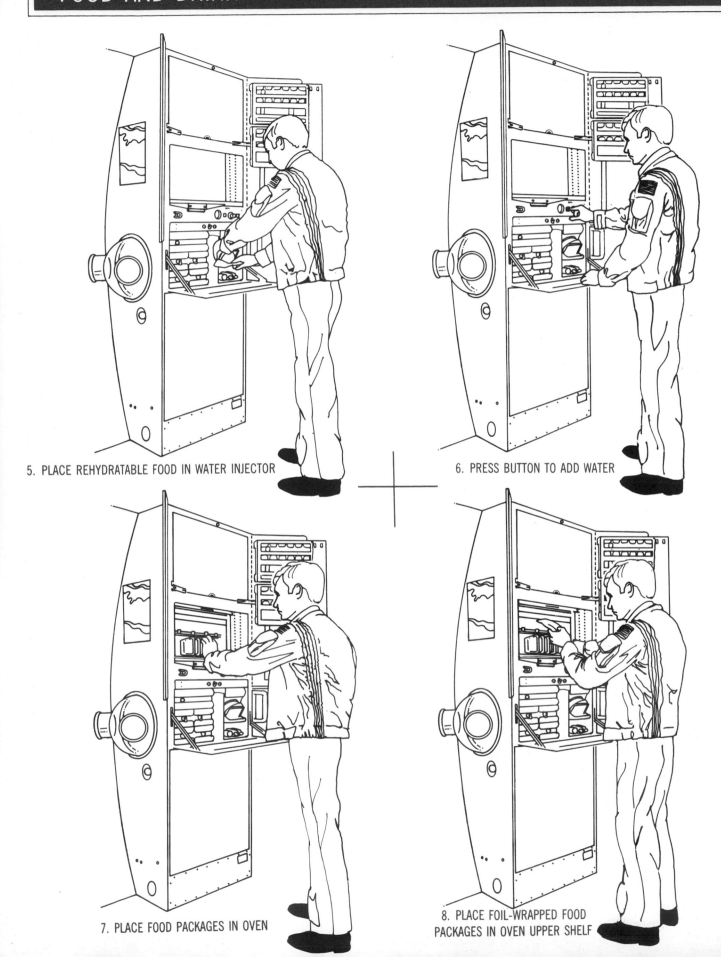

5. PLACE REHYDRATABLE FOOD IN WATER INJECTOR

6. PRESS BUTTON TO ADD WATER

7. PLACE FOOD PACKAGES IN OVEN

8. PLACE FOIL-WRAPPED FOOD
PACKAGES IN OVEN UPPER SHELF

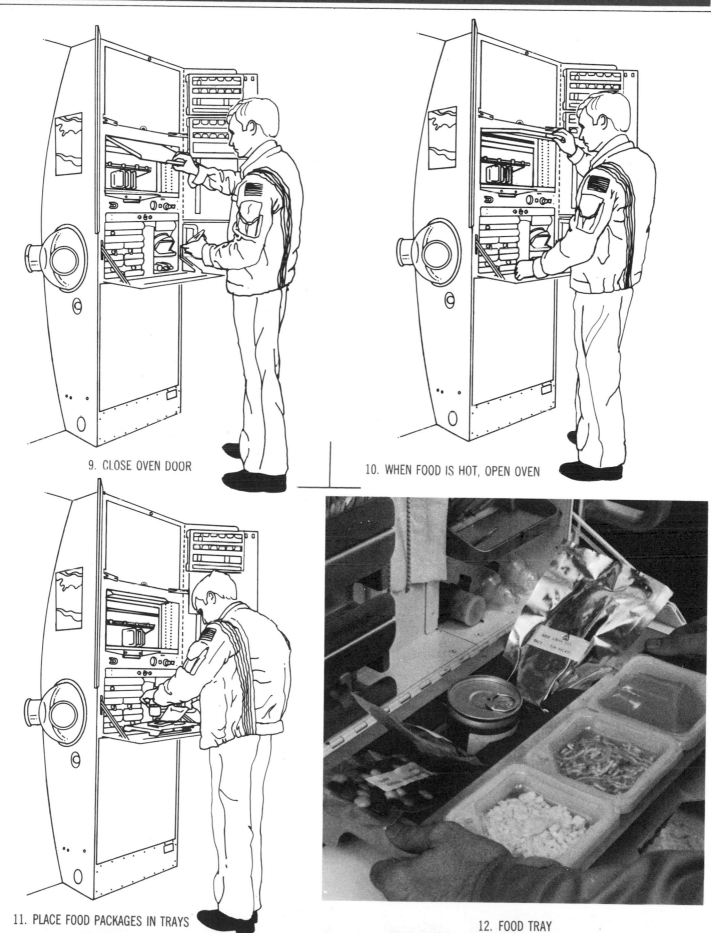

9. CLOSE OVEN DOOR

10. WHEN FOOD IS HOT, OPEN OVEN

11. PLACE FOOD PACKAGES IN TRAYS

12. FOOD TRAY

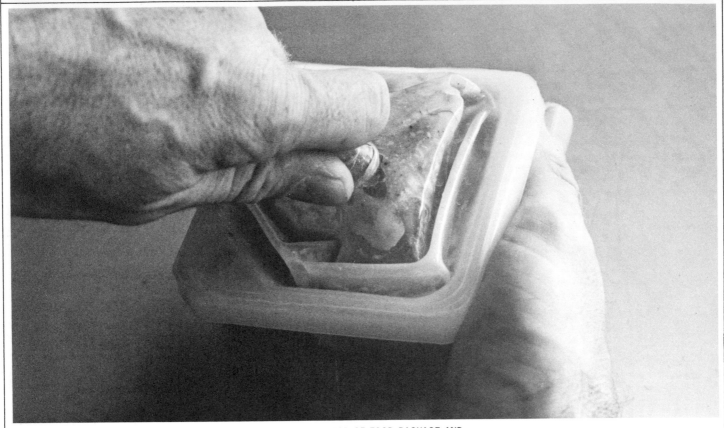

13. PULL UP TOP OF FOOD PACKAGE AND

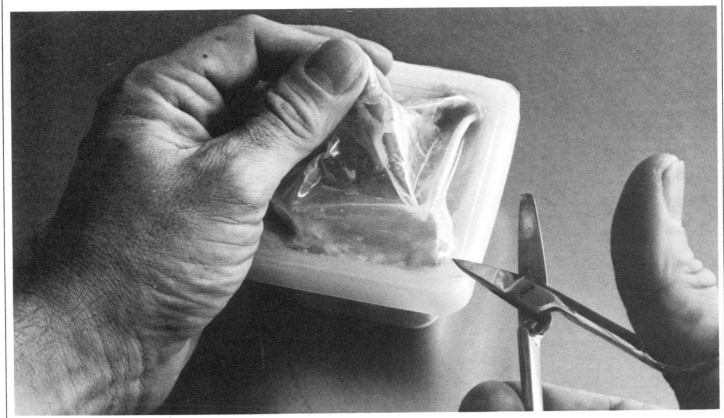

14. CUT OPEN

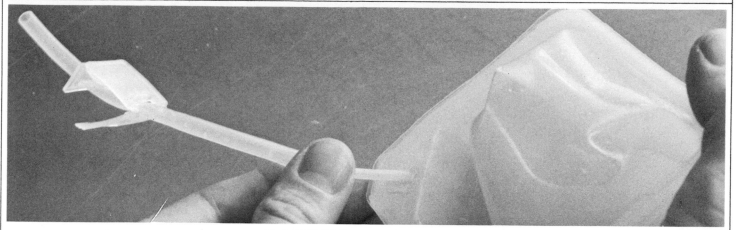

15. INSERT STRAW IN DRINK CONTAINER

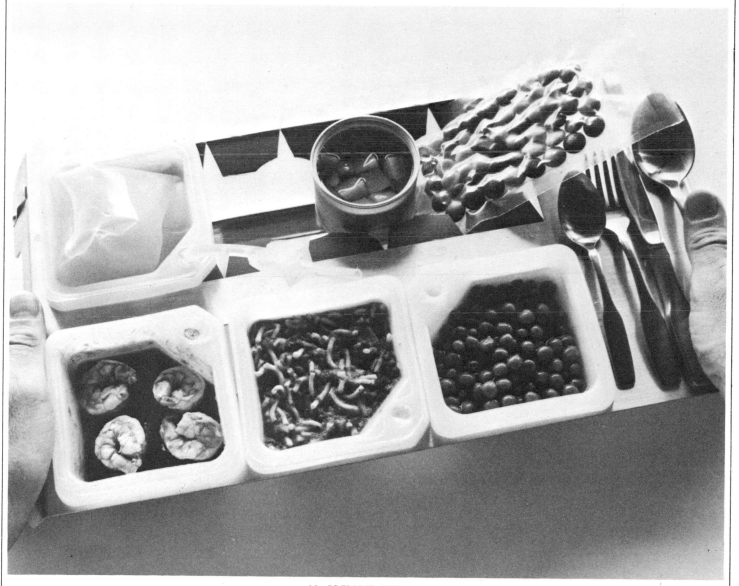

16. PREPARED MEAL

Sleeping accommodations aboard the Shuttle have been designed for comfort and privacy. The crew compartment on mid-deck has provisions for four **sleep stations**. (Crews of five or more usually sleep in shifts.) There are three horizontal bunks and one vertical bunk; remember, in microgravity up and down are relative terms. In fact, the bottom horizontal bunk faces the floor—so you sleep on the "ceiling" of the compartment.

The bunks are more than 6 feet (1.75 meters) long and 30 inches (.75 meter) wide. Each consists of a **sleeping bag** similar to those used in the Apollo spacecraft attached to a padded board. In weightlessness, the sleeping bag (also called a **sleep restraint**) holds you against the hard board with just enough pressure to create the illusion of sleeping on a comfortable mattress.

Your schedule allows 45 minutes to prepare for bed. You probably won't need that much time, but after a busy day some time to reflect and relax will be appreciated.

Store your shoes and outer clothing in your personal stowage container. The sleep restraint has a full-length zipper in front and an elastic waistband. The horizontal bunks are in compartments with movable panels for privacy; the vertical bunk has a fabric curtain. If all seven crew members sleep at the same time, three sleep restraints can be attached to the storage lockers on the mid-deck forward bulkhead.

You have a sleeping mask and earplugs if you need them—a spacecraft is never completely quiet, and it seems there is always a light on somewhere. When the entire crew sleeps at once, at least one person must wear a communication carrier to hear ground calls and on-board caution and warning alarms. Eight hours are set aside for sleep, and, in addition to the 45 minutes to prepare for bed, you are allotted 45 minutes when you wake for personal hygiene.

BUNK LOCATIONS

ORBITER MID DECK

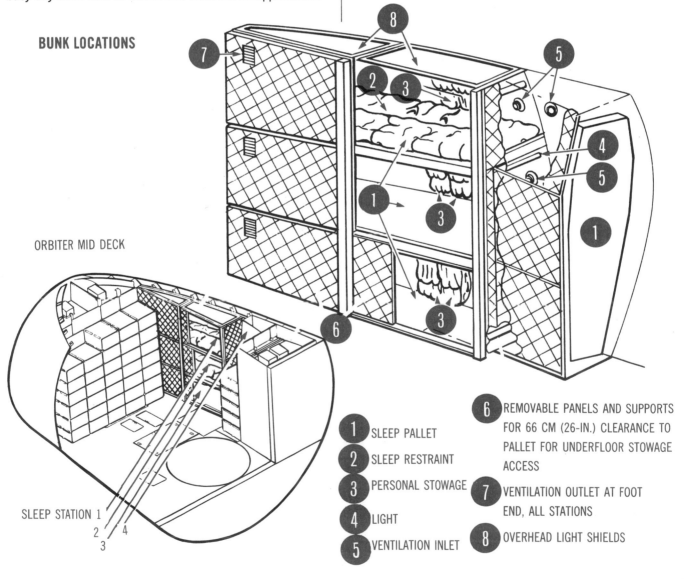

SLEEP STATION 1
2 4
3

1 SLEEP PALLET

2 SLEEP RESTRAINT

3 PERSONAL STOWAGE

4 LIGHT

5 VENTILATION INLET

6 REMOVABLE PANELS AND SUPPORTS FOR 66 CM (26-IN.) CLEARANCE TO PALLET FOR UNDERFLOOR STOWAGE ACCESS

7 VENTILATION OUTLET AT FOOT END, ALL STATIONS

8 OVERHEAD LIGHT SHIELDS

Keeping clean and properly disposing of garbage and body wastes are especially important in a closed-in area such as the crew compartment. In such a small volume, trash and waste materials would quickly build up and create a health hazard if not contained.

The Orbiter's toilet can be used by both men and women. It is in a closet in the mid-deck, next to the entry hatch. Because you're in a microgravity environment, there are some significant differences between the Shuttle's toilet and commodes on earth.

To use the toilet—or **waste collection system** (WCS), as it is called—first open the WCS compartment door and extend the pair of privacy curtains. To the right of the commode there's an operating handle and a control panel. When the handle is in the FORWARD position, the toilet **gate valve** opens, air is drawn through the toilet, and a set of vanes in the toilet begin rotating. In microgravity, airflow through the toilet and urinal moves wastes through the system. The moving vanes comprise an assembly called the **slinger**. The slinger shreds solid waste and deposits it in a thin layer on the walls of the commode chamber. Airflow makes sure the waste is drawn into the rotating blades. When the handle is in the OFF position, the gate valve closes and the WCS **vent valve** opens, exposing the commode chamber to the vacuum of space. This dries the feces deposited on the chamber's walls. The toilet can be used up to four times an hour.

Liquid waste is conveyed from the urinal to a waste-water tank. As with the commode, airflow moves the waste material through the system. Waste water is periodically dumped overboard.

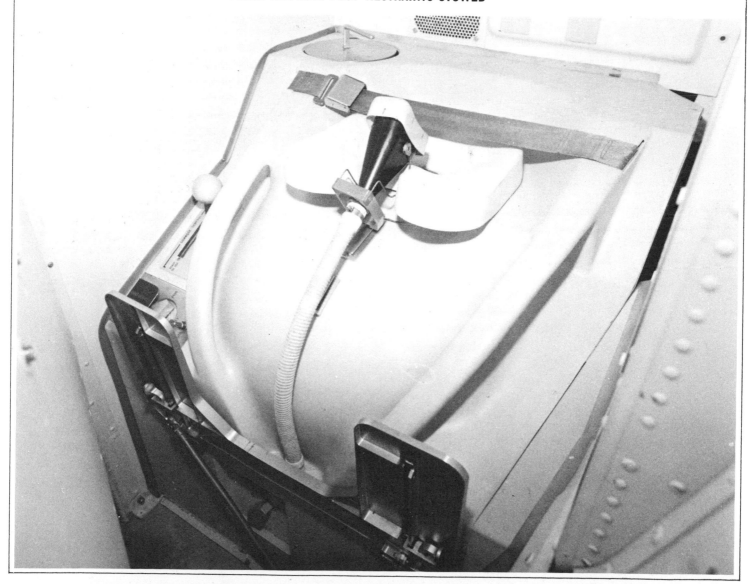

TOILET SHOWING FOOT RESTRAINTS STOWED

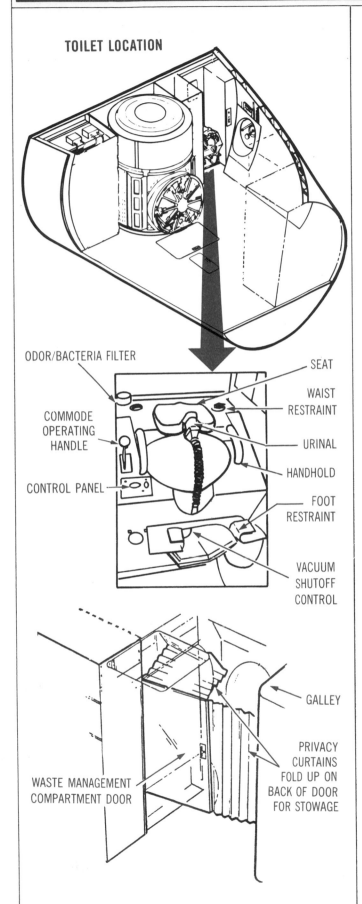

TOILET LOCATION

ODOR/BACTERIA FILTER

COMMODE OPERATING HANDLE

CONTROL PANEL

SEAT

WAIST RESTRAINT

URINAL

HANDHOLD

FOOT RESTRAINT

VACUUM SHUTOFF CONTROL

WASTE MANAGEMENT COMPARTMENT DOOR

GALLEY

PRIVACY CURTAINS FOLD UP ON BACK OF DOOR FOR STOWAGE

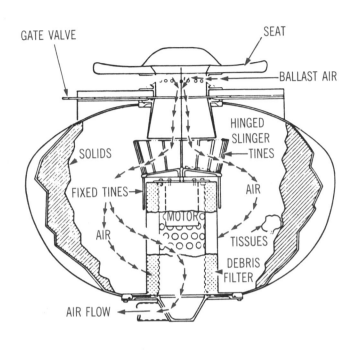

GATE VALVE

SEAT

BALLAST AIR

HINGED SLINGER TINES

SOLIDS

AIR

FIXED TINES

MOTOR

AIR

TISSUES

DEBRIS FILTER

AIR FLOW

COMMODE OPERATION

PULLING UP GATE VALVE CONTROL ACTIVATES SLINGER MOTOR AS SLINGER REACHES OPERATIONAL SPEED (1500 RPM), HINGED SLINGER TINES UNFOLD OUTWARD. PUSHING GATE VALVE CONTROL FORWARD OPENS GATE VALVE FOR COMMODE USE. FECES ENTERS COMMODE THROUGH SEAT OPENING, DRAWN IN BY BALLAST AIR FLOW. SLINGER TINES SHRED FECES AND DEPOSIT IT IN THIN LAYER ON COMMODE WALLS. TISSUES MOVE UP OVER SLINGER TINES AND SETTLE AT BOTTOM OF COLLECTOR. BALLAST AIR PASSES THROUGH DEBRIS FILTER AND HYDROPHOBIC FILTER TO FAN SEPARATORS.

FOR EMESIS DISPOSAL, FECAL/EMESIS SELECTOR SWITCH ON WCS CONTROL PANEL IS MOVED TO EMESIS POSITION, ROTATIONAL SPEED OF SLINGER TINES IS SLOWED AND TINES DO NOT UNFOLD. THIS ALLOWS UNOBSTRUCTED PASSAGE OF FECAL/VOMITUS BAG INTO COMMODE.

Trash and garbage are stowed until landing. Place dry trash in the bags provided around the crew compartment. Wet trash and garbage is placed in a special locker beneath the mid-deck floor. A circular opening in the floor allows you to put trash in the locker.

Your **personal-hygiene kit** contains small items you need for oral hygiene, hair care, shaving, and personal comfort. The kit contains a toothbrush, toothpaste, dental floss, nail clippers, soap, a comb, a brush, antichap lipstick, skin lotion, and stick deodorant. For male crew members, it also contains a tube of shaving cream and a safety razor, or a wind-up shaver.

A **personal-hygiene station** is on the side of the galley facing the WCS. It has a light, a mirror, and a hand-washing enclosure. There's no shower—you'll have to wash using a washcloth at the personal-hygiene station. Each crew member has seven washcloths and three towels.

HANDWASHING STATION

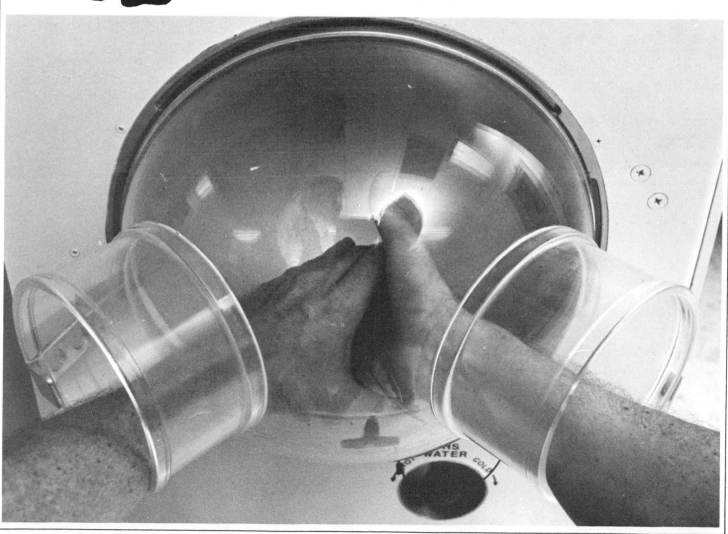

Your clothing has been designed to be comfortable and functional. Since the crew compartment has an atmosphere like the earth's, you don't have to wear a bulky pressure suit except for extravehicular activities, or "space-walks" (see 3.9). Instead, the Crew Systems Division issued you a smart-looking outfit consisting of a cobalt-blue jacket and trousers with a short-sleeve navy-blue cotton knit shirt. Stocks of these items, along with your underwear, socks, footwear, and gloves, are maintained in standard sizes and are issued "off the shelf." All clothing, except underwear, is the same for both sexes.

The waist-length jacket is fitted by chest size and sleeve length. It is lined, made from soft cotton, and has a full-length zipper in its front. Expansion pleats in the shoulders make it easier for you to move and flex while wearing the jacket. The trousers are fitted by waist and inseam measurements. Like the jacket, they are made of soft cotton.

Both safety and convenience were considered when the outfits were designed. The clothing is loose enough to be comfortable without being sloppy—loose clothing can accidentally catch on critical switches as you brush by them. Pockets cover much of the flight suit's exterior. They are closable, either with Velcro or zippers, so you can use them to secure such small items as scissors, pens, sunglasses, and data books. (If not secured, small items will float around in zero-g and become at least a nuisance, at worst a hazard.) The clothes have all been soaked with a fireproofing chemical.

INFLIGHT COVERALLS

Clothing

Underwear	1 set per day
One-g footwear	1 pair per flight
Jacket	1 per flight
Trousers	1 pair per 7 days plus 1 spare per flight
Shirt	1 per 3 days
Gloves	1 pair per flight
In-flight footwear	1 pair per flight
Brassiere (female personnel only)	1 per day

Personal Items

Felt-tip and pressurized pens	Variable
Mechanical pencils	Variable
Sunglasses	1 pair per flight
Swiss Army knife	1 per flight
Surgical scissors	1 per flight
Chronograph (watch)	1 per flight
Sleeping mask and earplugs	1 per flight

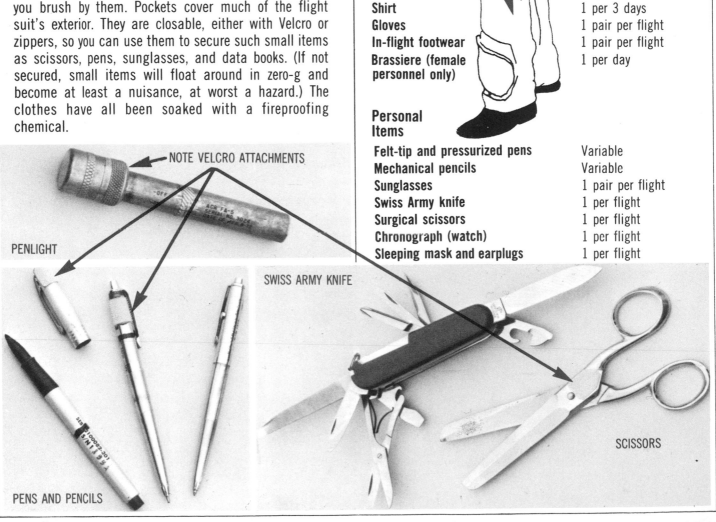

NOTE VELCRO ATTACHMENTS

PENLIGHT

PENS AND PENCILS

SWISS ARMY KNIFE

SCISSORS

The uniqueness of the Shuttle lies not only in its reusability and carrying capacity, but also in the variety of jobs it can do in orbit. The Shuttle brings payloads to space, and it can service them or collect them for return. Shuttle missions are both scientific and practical. The research opportunities include biological studies of humans or other animals or plants in the microgravity of space. Astronomers can look at the sun, stars and galaxies with detectors that would be useless under the earth's atmosphere. The Shuttle will also probe the earth's environment— oceans, air, land use, and weather. Studies of how materials react in microgravity may produce some startling ideas for industry.

The practical payloads will monitor the earth's environment continuously, relay communications services—television, telephone, and computer data—and service any previously launched satellites or future space stations.

Work is what the Shuttle is about. This section will give you information and instructions for crew operations on basic Shuttle missions.

Controls in the back of the flight deck make up the **aft crew station**. The station comprises the necessary controls to manage the Orbiter and its payload in space. From here, you can maneuver the craft, operate the manipulator arm, and monitor payloads.

If you face the aft portion of the flight deck, you'll see the controls arranged in a shallow U. There are two overhead windows and two windows on the aft bulkhead for looking into the cargo bay. In the left branch of the U is the mission station. From here, the mission specialist manages the flight. He has a computer keyboard and spacecraft system monitors. On the opposite side of the cabin, the payload specialist has monitors and controls for each payload.

Between these two work locations, on the back wall, are the pilot's and payload handler's stations. Using the left-hand set of controls, the pilot maneuvers the Orbiter for docking or payload handling. On the pilot's right, the payload handler operates the manipulator arm and the television system.

In orbit, the aft crew station becomes the control center for your work with payloads.

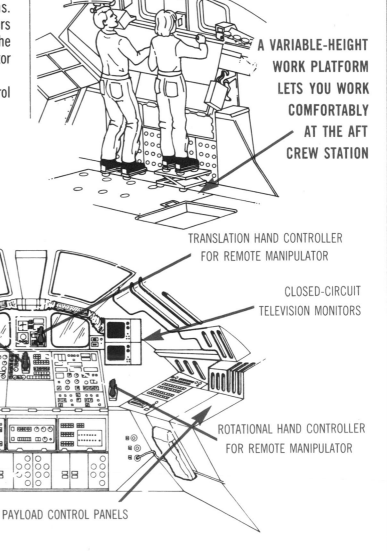

A VARIABLE-HEIGHT WORK PLATFORM LETS YOU WORK COMFORTABLY AT THE AFT CREW STATION

ORBITER ROTATIONAL HAND CONTROLLER

ORBITER TRANSLATION HAND CONTROLLER

TRANSLATION HAND CONTROLLER FOR REMOTE MANIPULATOR

CLOSED-CIRCUIT TELEVISION MONITORS

ROTATIONAL HAND CONTROLLER FOR REMOTE MANIPULATOR

PAYLOAD CONTROL PANELS

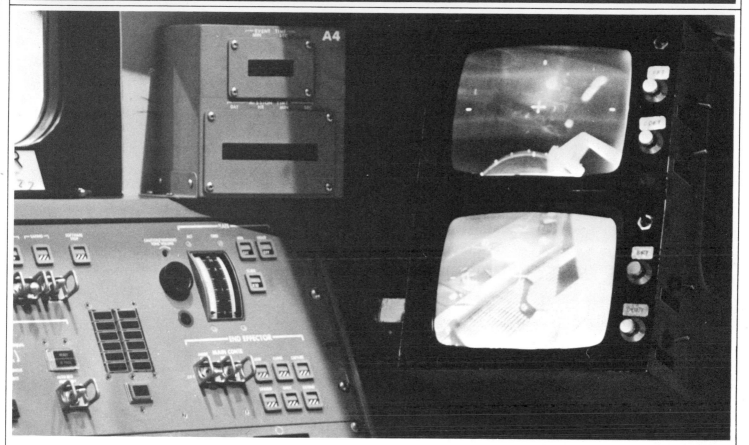

AT WORK IN AFT CREW STATION

The **remote manipulator system** (RMS) is a mechanical arm which moves objects in and out of the cargo bay in space. It has three joints: **shoulder, elbow, and wrist**. The shoulder is where the 50-foot (15-meter) arm attaches to the Orbiter. Like your own arm, the RMS has an elbow in the middle and a wrist for its hand, or **end effector**.

Inside the Orbiter, on the right side of the aft crew station, are the controls for the arm. You'll also need to use the controls on the station's left side to maneuver the spacecraft into the proper position. Once you're in position, turn on the cargo-bay floodlights, using the switches on panel A7.

There are television cameras on the arm's elbow and end effector. These allow you to observe all the movements made by the arm so you don't damage the Orbiter or its payload. The video controls are also on panel A7, under the heading TV. Select the camera you want. If the video is also to be transmitted to the ground, the TV DOWNLINK switch should be in the ENABLE position. Two black-and-white monitors are located to your right. These enable you to simultaneously see the outputs from two cameras, such as those on the RMS. The CAMERA COMMAND switches let you focus, control image brightness, and zoom in and out.

REMOTE MANIPULATOR SYSTEM AND CONTROLS

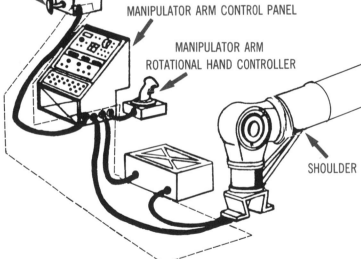

ORBITER ROTATIONAL HAND CONTROLLER

MANIPULATOR ARM CONTROL PANEL

MANIPULATOR ARM
ROTATIONAL HAND CONTROLLER

SHOULDER

RMS control can be either manual or automatic. To move the arm manually, first locate the arm controls in front of the aft wall. Between the windows, on panel A2, you'll find the arm-translation controller. Moving this knob up or down, in or out, or sideways causes a like motion in the arm. To your right, there's a hand controller for arm rotation. On panel A8, turn the MODE switch to the ORB UNL (Orbiter, unloaded) position. Now, translation movements by the arm will be parallel to the Orbiter's axes. On the JOINT switch (also on panel A8), select SHOULDER, PITCH. Push up on the translation controller. The arm lifts up while the end effector remains parallel to the Orbiter's long axis. Now select SHOULDER, YAW, and move the translation controller from side to side. Be careful when moving it to the left—don't hit any payloads

in the cargo bay. Using a similar procedure, try the rotation controller. With practice, you'll be able to coordinate translation and rotation movements to reach payloads in the bay or floating freely in space.

The end effector has a three-wire capture mechanism that grips the standard docking target, or **grapple**, on all free-flying payloads. When the grapple probe is in the triangle formed by the three wires, a ring attached to one end of each wire rotates. This causes the triangle to shrink until the wires are against the probe. The wire assembly then withdraws into the end effector and pulls the grapple tight against it. You can then maneuver the arm and payload, but be sure to place the MODE switch in the ORB LD (Orbiter, loaded) position.

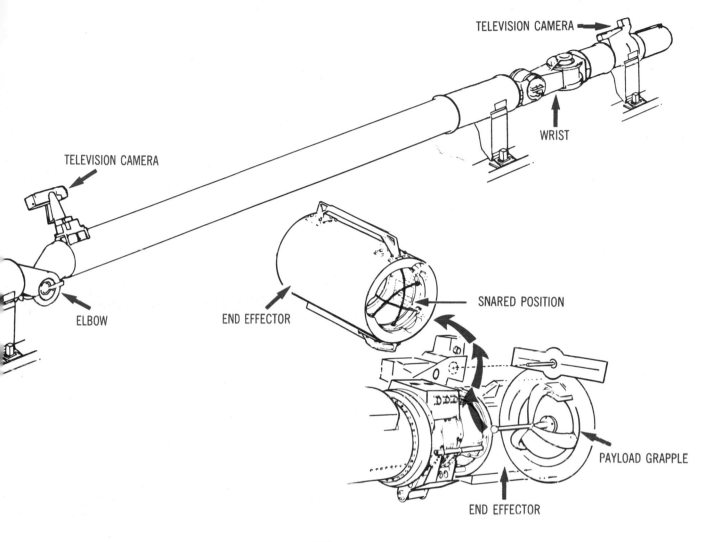

TELEVISION CAMERA

TELEVISION CAMERA

WRIST

ELBOW

END EFFECTOR

SNARED POSITION

PAYLOAD GRAPPLE

END EFFECTOR

GRAPPLE CAPTURE

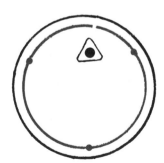

PAYLOAD GRAPPLE INSIDE
OPEN END OF END EFFECTOR:
WIRES STORED.

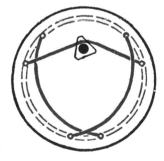

END EFFECTOR RING BEGINS
TO ROTATE; WIRES BEGIN TO
CLOSE ONTO PAYLOAD GRAPPLE

END EFFECTOR RING FULLY ROTATED;
WIRES CLOSED ON PAYLOAD GRAPPLE,
CENTERING IT AND CAPTURING PAYLOAD

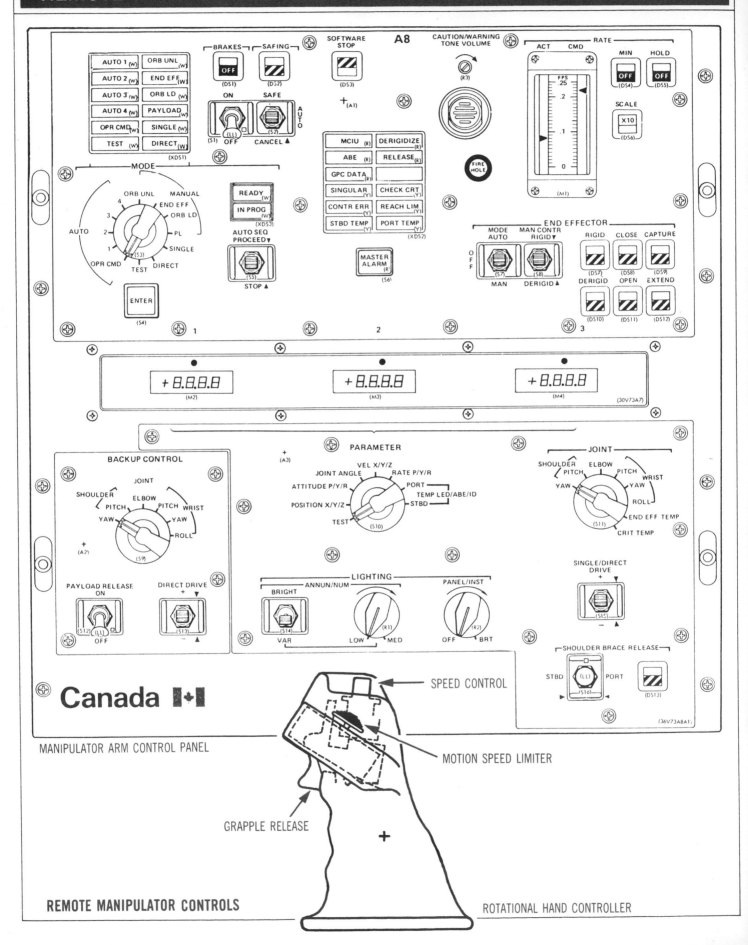

MANIPULATOR ARM CONTROL PANEL

Canada 🍁

SPEED CONTROL

MOTION SPEED LIMITER

GRAPPLE RELEASE

REMOTE MANIPULATOR CONTROLS

ROTATIONAL HAND CONTROLLER

Many of the Shuttle's payloads operate in geostationary orbits 22,300 miles (35,900 kilometers) above the earth's surface where it takes 24 hours to complete one orbit. A satellite in such an orbit over the equator remains fixed above the same point on the ground, providing a stable platform for communications relays. The Orbiter, however, is limited to orbits below 690 miles (1,100 kilometers). To bridge the gap between the Shuttle's capabilities and the desired higher orbits, **upper-stage rocket motors** are attached to the payloads. Three solid-propellant motors have been developed as part of the Space Transportation System.

The McDonnell Douglas Astronautics Company designed two of the motors. They call them **payload assist modules**; NASA refers to them as **spinning solid upper stages**. Both motors are sold by McDonnell Douglas for non-NASA payloads, so most people use the manufacturer's name, abbreviated PAM.

The smaller of the PAMs was designed to accommodate payloads similar to those previously launched by the Delta launch vehicle, so it is called PAM-D. It can place up to 2,000 pounds (900 kilograms) into a geosynchronous transfer orbit. The other PAM is for heavier payloads, up to 4,400 pounds (2,000 kilograms). Payloads of this weight used to be launched by the Atlas Centaur, so this upper stage is called PAM-A. Both PAMs are single-stage; that is, each contains a single rocket motor.

The operating sequence for both versions of the PAM is the same. When the Shuttle lifts off, the payload and its attached PAM are held in a cradle in the cargo bay. The base of the PAM is attached to a spin table. Prior to release, electric motors in the table spin the PAM/payload assembly about its long axis. This spin continues after the assembly is released from the cradle and cancels out off-balances in the motor's thrust caused by irregular propellant burning. PAM ignition occurs 45 minutes after release to separate it from the Orbiter. After PAM burnout, the expended motor casing separates from the payload, which is finally placed in a circular geosynchronous orbit by a self-contained **apogee kick motor.**

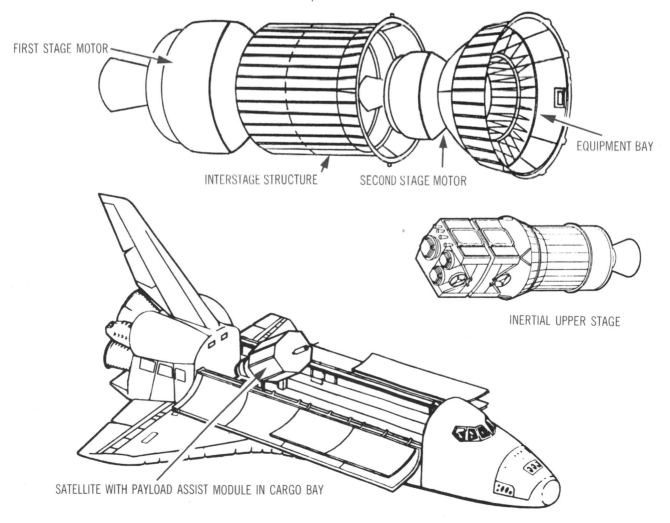

FIRST STAGE MOTOR

INTERSTAGE STRUCTURE   SECOND STAGE MOTOR   EQUIPMENT BAY

INERTIAL UPPER STAGE

SATELLITE WITH PAYLOAD ASSIST MODULE IN CARGO BAY

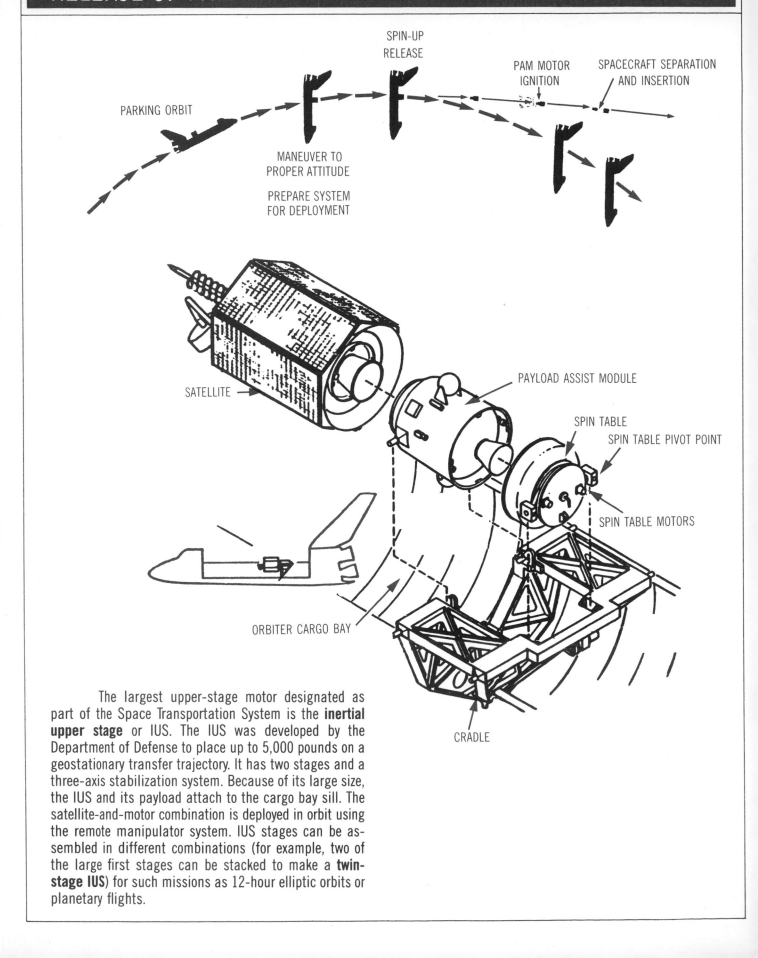

SPIN-UP RELEASE

PAM MOTOR IGNITION

SPACECRAFT SEPARATION AND INSERTION

PARKING ORBIT

MANEUVER TO PROPER ATTITUDE

PREPARE SYSTEM FOR DEPLOYMENT

SATELLITE

PAYLOAD ASSIST MODULE

SPIN TABLE

SPIN TABLE PIVOT POINT

SPIN TABLE MOTORS

ORBITER CARGO BAY

CRADLE

The largest upper-stage motor designated as part of the Space Transportation System is the **inertial upper stage** or IUS. The IUS was developed by the Department of Defense to place up to 5,000 pounds on a geostationary transfer trajectory. It has two stages and a three-axis stabilization system. Because of its large size, the IUS and its payload attach to the cargo bay sill. The satellite-and-motor combination is deployed in orbit using the remote manipulator system. IUS stages can be assembled in different combinations (for example, two of the large first stages can be stacked to make a **twin-stage IUS**) for such missions as 12-hour elliptic orbits or planetary flights.

Soviet Cosmonaut Alexei Leonov performed the first **extravehicular activity** (EVA, also popularly referred to as a **spacewalk**) in March, 1965. Less than three months later, Astronaut Edward H. White II became the first American to equal that feat. Leonov floated alongside Voskhod-2 for 12 minutes. White's EVA lasted 22 minutes as he "strolled" across North America. These brief excursions tested man's ability to venture outside an orbiting spacecraft protected only by a pressurized spacesuit. Subsequent EVAs showed that space travelers could perform useful work while floating outside their spacecraft or walking on the surface of the Moon.

Perhaps the most dramatic spacewalk thus far occurred on June 7, 1973, when Charles Conrad and Joseph Kerwin saved the crippled Skylab space station and salvaged a multi-billion-dollar program. During launch the micrometeoroid/thermal shield wrapped around the Orbital Workshop tore loose, taking one of the vehicle's solar panels with it. Debris from the shield jammed the other panel and prevented it from unfolding. With the remaining panel folded, Skylab was without electric power and the entire project was in peril. Conrad and Kerwin, working outside the Workshop, were able to cut through the debris and deploy the solar panel, thus saving Skylab. None of the remaining nine EVAs performed by the three Skylab crews was as spectacular, but each accomplished some important task—retrieving a film cassette, repairing antennas, cleaning experiment optics, etc.

On Space Shuttle flights the tradition of using EVAs to expand man's capabilities in space continues. Shuttle EVAs fall into three categories:
1. **Planned**—EVAs planned prior to launch to fulfill mission objectives.
2. **Unscheduled**—EVAs that are not planned, but that become necessary during the flight for payload operation success.
3. **Contingency**—Emergency EVAs required to save the Orbiter and/or its crew.

Each Shuttle mission carries enough equipment and consumables for three two-person, 6-hour EVAs. Two of these are available for payload operations. The third is reserved for handling an in-flight emergency.

The major pieces of personal extravehicular equipment are the **spacesuit** (more properly referred to as the **extravehicular mobility unit**, or EMU) and the **manned maneuvering unit** (MMU).

The Shuttle's EMU features many significant departures from previous spacesuits. The EMU costs less than earlier suits and is more flexible. Tailor-made customized suits are things of the past. Shuttle EMUs are manufactured in several standard sizes; you adjust straps inside the suit to make it fit. Each suit has a 15-year life expectancy. Your EMU consists of three assemblies: the **upper torso**, the **lower torso** or trousers, and the **portable life-support system**.

The upper torso and trousers separate into two units. A connecting ring around the waist joins them, eliminating the need for any zippers in the suit. The chest portion of the upper torso is a rigid assembly made from aluminum. Ring joints connect the arms to the torso. **Constant-volume joints** in the shoulders and elbows make it easier to move in this suit than in past designs. A constant-volume joint resembles a bellows. As the joint bends, one side contracts while the other expands, keeping the interior volume, and hence the pressure, constant. Without this type of joint, flexing your elbow would be like bending an over-inflated balloon. Constant-volume joints are also used in the trousers at the hips, knees, and ankles.

Another new feature of the Shuttle EMU is the choice of polyurethane for the suit's pressure bladder. Pressure bladders for the Apollo suit were made from latex rubber and were glued and taped together. Seams in the newer polyurethane bladders are heat-sealed to prevent small leaks from developing with use.

The portable life-support system is contained in a backpack that is permanently attached to the upper torso. All connections between the life-support unit and the spacesuit are inside the suit, eliminating the external hoses and connections seen on past suits. The unit contains enough oxygen and electric power for 7 hours. This allows 15 minutes to check the suit after you have donned it, 6 hours for extravehicular activity, 15 minutes to take off the suit, and 30 minutes for reserve. In addition, an emergency 30-minute supply of oxygen is contained in the secondary oxygen pack. The life-support system can be recharged between spacewalks using Orbiter systems.

A chest-mounted microcomputer with a **light-emitting diode** (LED) display provides you with constant status checks of oxygen and battery power. Additionally, if there is a suit malfunction, the microcomputer provides a warning and specifies the appropriate corrective action. The microcomputer is housed in the **display and control module**, which also contains the electrical and mechanical controls required to operate the EMU.

Beneath the suit, you wear a **cooling and ventilation garment** similar to the ones worn by Apollo and Skylab astronauts. The garment is a one-piece affair made from Spandex mesh. Plastic tubing is woven into the mesh, and cool water from the life-support backpack circulates through the tubing to remove excess body heat. Air ducts attached to the garment provide ventilation to your limbs. This is a new feature; Apollo EMUs had ventilation tubes built into the space suits rather than the cooling garment. Pressure gloves, an Apollo-style plastic bubble helmet, and a snap-on visor complete the garment. The gloves, made in fifteen sizes, are similar to those worn on the lunar surface with molded rubber finger caps to permit tactile response. The visor on the helmet provides protection from micrometeoroids and ultraviolet radiation.

The Space Shuttle EMU has been carefully designed to allow you to put it on and take it off by yourself. Putting on the suit takes 5 minutes. Donning an Apollo spacesuit required the assistance of another person and took at least 30 minutes.

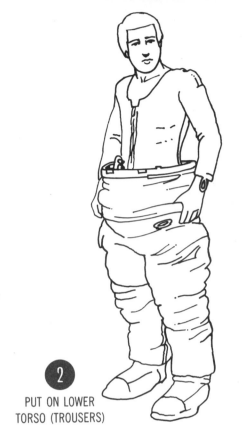

2

PUT ON LOWER
TORSO (TROUSERS)

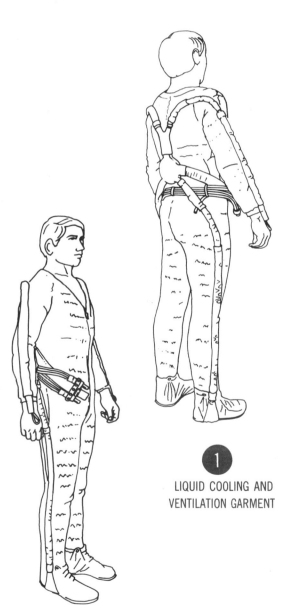

1

LIQUID COOLING AND
VENTILATION GARMENT

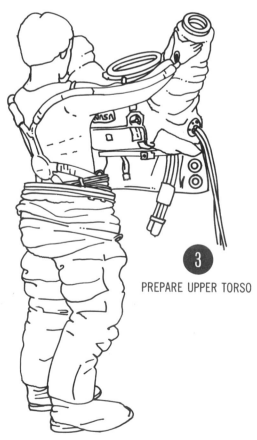

3

PREPARE UPPER TORSO

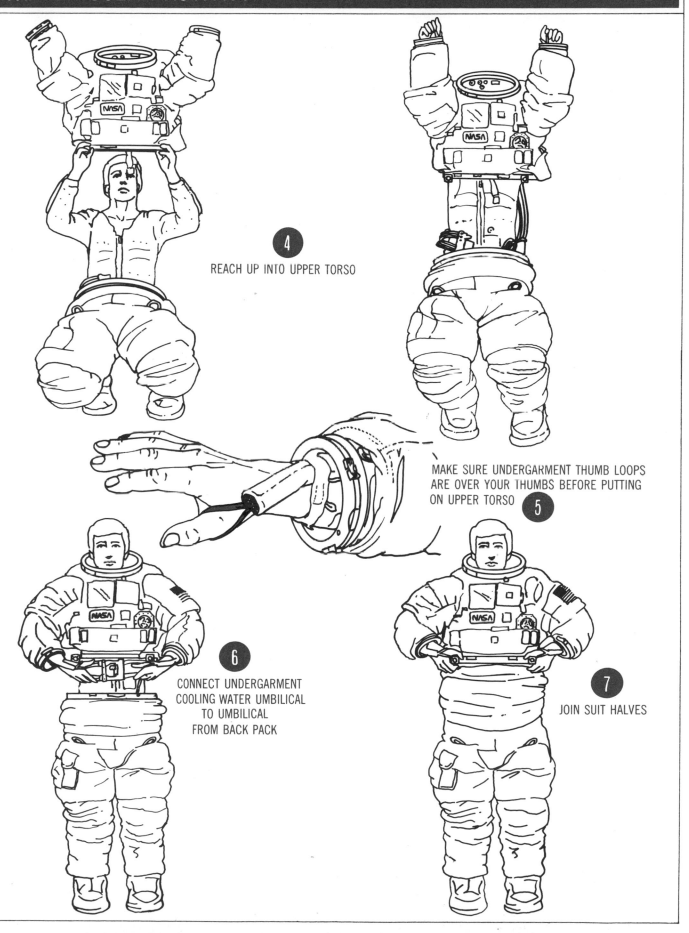

4

REACH UP INTO UPPER TORSO

MAKE SURE UNDERGARMENT THUMB LOOPS
ARE OVER YOUR THUMBS BEFORE PUTTING
ON UPPER TORSO

5

6

CONNECT UNDERGARMENT
COOLING WATER UMBILICAL
TO UMBILICAL
FROM BACK PACK

7

JOIN SUIT HALVES

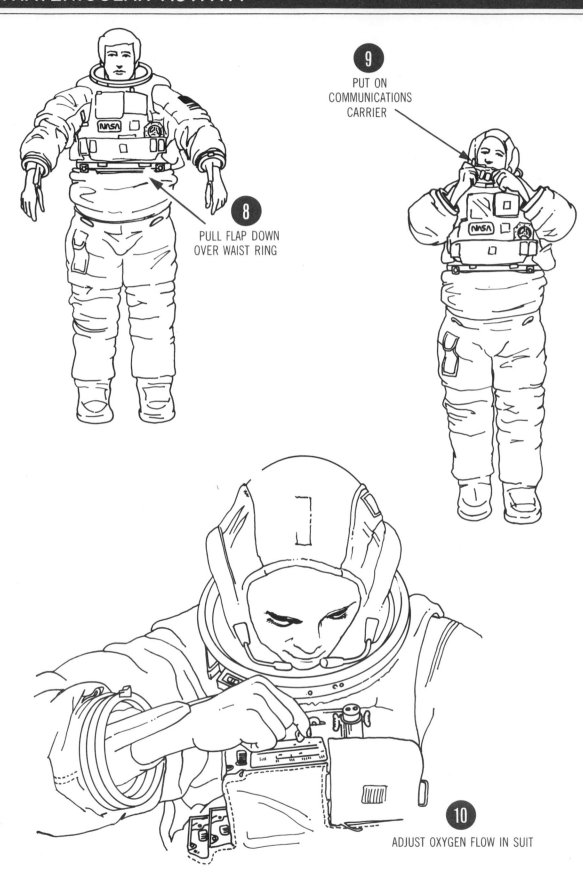

8 PULL FLAP DOWN OVER WAIST RING

9 PUT ON COMMUNICATIONS CARRIER

10 ADJUST OXYGEN FLOW IN SUIT

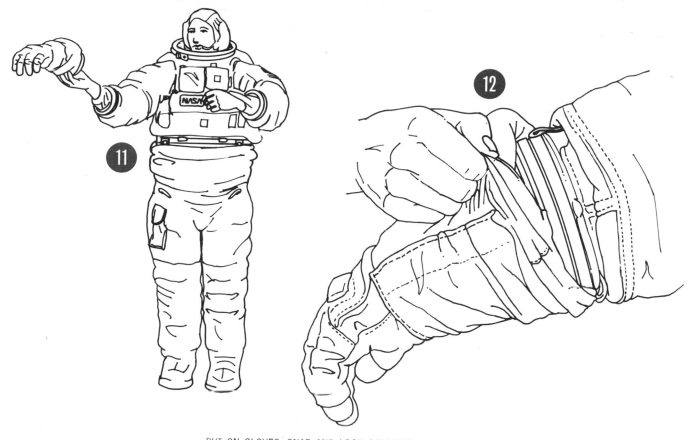

PUT ON GLOVES, SNAP AND LOCK CONNECTING RINGS

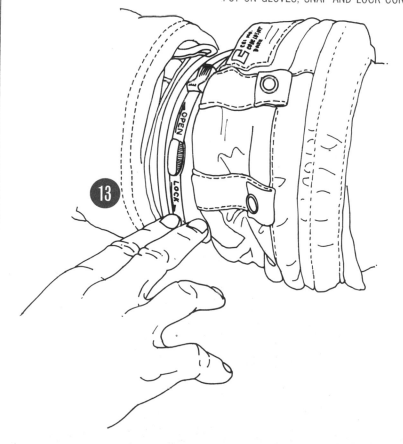

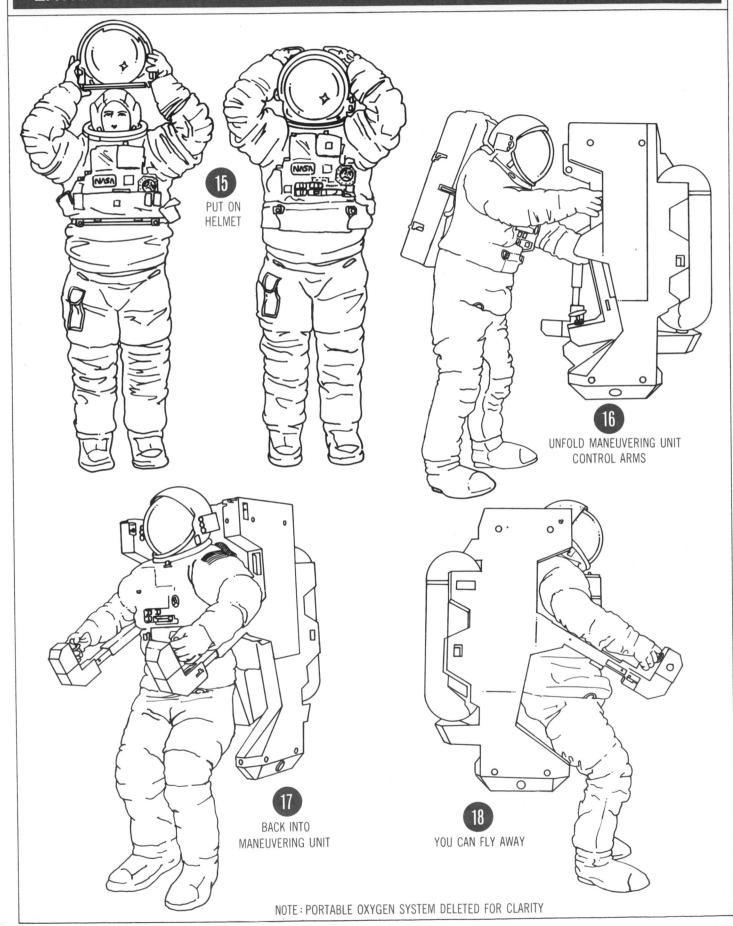

15 PUT ON HELMET

16 UNFOLD MANEUVERING UNIT CONTROL ARMS

17 BACK INTO MANEUVERING UNIT

18 YOU CAN FLY AWAY

NOTE: PORTABLE OXYGEN SYSTEM DELETED FOR CLARITY

The manned maneuvering unit, or MMU, allows you to move from the Orbiter to other orbiting spacecraft. The MMU is a self-contained backpack that latches onto your spacesuit. Two latches connect the MMU to your spacesuit's life-support backpack. The latches allow you to don and doff the maneuvering unit by yourself. A fiber-optics cable links the MMU to the suit's display and control module on your chest. This allows you to monitor the MMU's operation the same way you monitor that of the spacesuit. Readouts show propellant quantity, battery power level, and malfunctions along with the necessary corrective action. Normally, only one maneuvering unit is carried per flight, but another can be carried for an Orbiter rescue mission, or when flight plans require two. (When not in use, the unit is stored in the front of the cargo bay.)

Nitrogen gas propels the MMU. Twenty-four nozzles arranged around its exterior expel gas in spurts of 1.6 pounds of thrust (7 newtons) each. There is enough propellant for a total velocity change of about 66 feet (19 meters) per second. This is more than enough for you to fly around and inspect the Orbiter's exterior or fly between the Shuttle and another orbiting object.

Flight controls are on the ends of the MMU's arm rests. The right-hand control is for orientation. Up or down movement of the knob causes you to pitch up or down; left or right movement causes a left or right yaw, and rotating the handle rotates the unit. The left handle controls straight-line motion.

Preparation for your EVA begins 2½ hours ahead of time. The cabin atmosphere is 79% nitrogen and 21% oxygen at a pressure of 14.7 psi (100 kilopascals). The spacesuit is pressurized with pure oxygen at 4.1 psi (28 kilopascals). The suit's lower pressure represents the **partial pressure** (or content) of oxygen in the atmosphere and is sufficient to sustain life. However, there is one major problem—if you go directly from the oxygen-nitrogen cabin atmosphere into the pure-oxygen reduced-pressure environment in your suit, nitrogen gas dissolved in your blood will bubble out. The gas bubbles would collect in your joints and cause a condition known as **dysbarism**, or more popularly, the **bends**. This is, at the least, very painful and it can cripple or even kill. To prevent the bends, you must breathe pure oxygen for at least 2 hours before donning your spacesuit. This "washes" the nitrogen out of your body.

You have a **portable oxygen system** (POS) for this. Your POS can use either Orbiter-supplied oxygen (**dependent mode**) or its built-in supply (**independent or "walk-around" mode**). Normally, you'll use the dependent mode to save the internal supply for emergencies. The oxygen system has a face mask, an oxygen bottle, a

pressure regulator, and a lithium hydroxide cartridge. It also has an attachable hose for the dependent mode and a short hose with a scuba-style mouthpiece so you can continue prebreathing oxygen while putting on your space suit.

The lithium hydroxide cartridge is called the **recharge kit**. When you breathe out, your exhalation gases are routed through the recharge kit. Not all the oxygen you inhaled will have been converted to carbon dioxide, so when you exhale there is still quite a bit of oxygen present. (This is why mouth-to-mouth resuscitation is possible.) The lithium hydroxide removes the carbon dioxide, and the remaining gas (oxygen) is routed back to your mask. Oxygen is added as needed to replenish what is consumed.

About 30 minutes before your EVA is set to begin, switch from the face mask to the mouthpiece. Then, remove your coveralls and put on your liquid-cooling and ventilation garment and your urine-collection device. You are now ready to enter the airlock and don the EMU.

The **airlock**, a small cylindrical chamber, allows you to perform an EVA without depressurizing the entire crew compartment. It can be in the mid-deck, in the cargo bay attached to the crew compartment aft bulkhead, on a Spacelab transfer tunnel, or on a docking adapter. In all cases, it operates the same way and you exit from it into the cargo bay. Check the airlock's life-support system and close the entry hatch after you. Your spacesuit is mounted on a pipe frame on the airlock wall.

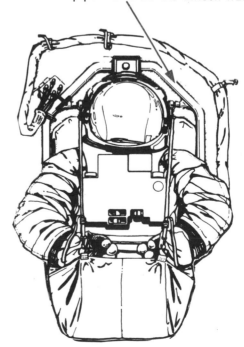

Check the EMU for any rips or tears and make sure you have all its parts. Once you're satisfied, put on the trousers or lower torso assembly. Next, wriggle up into the upper torso. As you put on the upper torso, make sure the loops on the cuffs of your undergarment are over your thumbs. Otherwise, the sleeves will ride up and you will be very uncomfortable.

Attach the liquid-cooling garment connector to the umbilical from the life-support backpack. You're now ready to connect the upper and lower torso assemblies. As you close the waist ring, you should feel and hear a series of faint clicks as the ring fasteners catch. Put on your **communications carrier** (popularly called the **Snoopy hat**) and connect it to the suit's communications umbilical. Set the sliding oxygen-control switch on your chestpack to PRESS and put on your helmet and gloves. In the PRESS position, your suit will build up to a pressure of 4 psi (28 kilopascals) above ambient conditions so you can check the garment for leaks.

Disconnect the suit from the wall mount. Turn to the airlock control panel and locate the AIRLOCK DEPRESS switch on panel AW 82A. Twist the knob to position 5. This reduces airlock pressure to 5 psi (34 kilopascals). Finally, turn it to its second position, 0. This depressurizes the airlock in about 3 minutes. At the same time, move the oxygen control on your chestpack to the EVA position. When the airlock pressure is less than 0.2 psi, (1.38 kilopascals) you can open the outer hatch and begin your EVA.

POSSIBLE AIRLOCK LOCATIONS

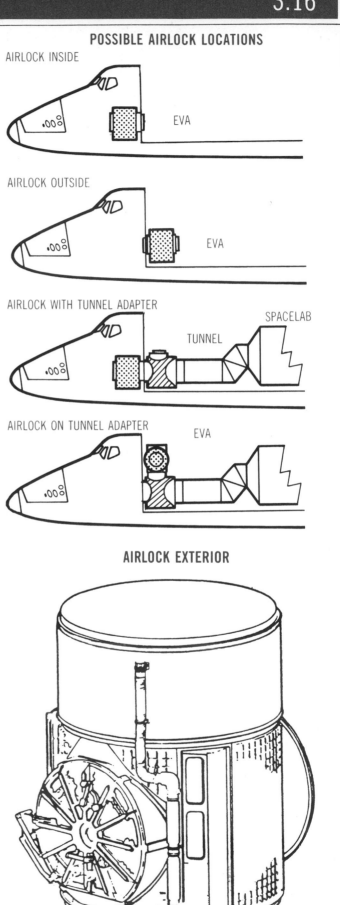

AIRLOCK INSIDE / EVA

AIRLOCK OUTSIDE / EVA

AIRLOCK WITH TUNNEL ADAPTER / TUNNEL / SPACELAB

AIRLOCK ON TUNNEL ADAPTER / EVA

AIRLOCK

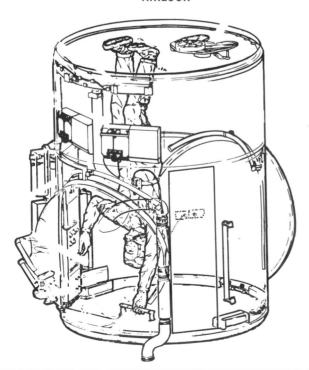

AIRLOCK EXTERIOR

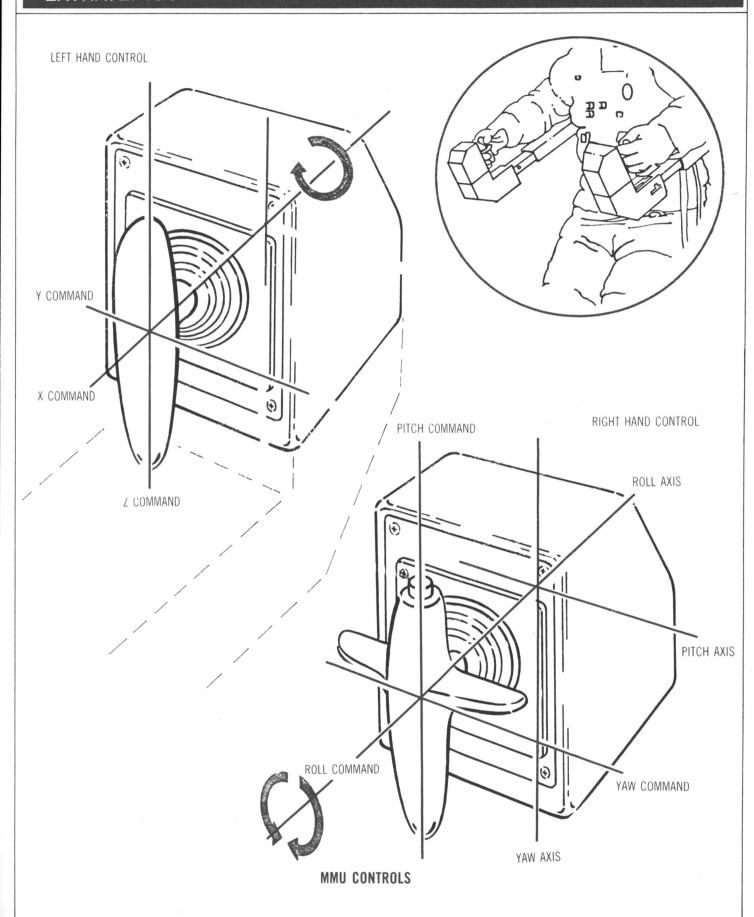

LEFT HAND CONTROL

Y COMMAND

X COMMAND

Z COMMAND

PITCH COMMAND

RIGHT HAND CONTROL

ROLL AXIS

PITCH AXIS

ROLL COMMAND

YAW COMMAND

YAW AXIS

MMU CONTROLS

One basic requirement for a successful mission is **communication**. You must be able to communicate with ground controllers and your fellow crew members in other parts of the Shuttle. An impressive assortment of hardware supports this requirement. Onboard equipment includes personal headsets, built-in intercoms, television cameras, and a teleprinter.

Your **headset** consists of an earphone, with clip, and a microphone. A cable connects the headset to a small control unit, which you clip to your clothes. The unit has a rocker switch that allows you to select either the onboard intercom or external communication. The intercom setting on the rocker switch is marked ICOM; the transmit position is marked XMIT PTT. There's a volume control on top of the box.

The cable extending from the bottom of your control unit connects (via an extension cord) to one of the eight **intercom** boxes in the crew compartment. Actually the box is more than just an intercom. Called an **audio terminal unit** (ATU), it allows you to communicate with the ground, another terminal in the Orbiter, or both. Communicating with another terminal in the crew compartment may seem unnecessary, but you'll appreciate it the first time you're working in the mid-deck and want to talk to someone on the flight deck.

You can select the **push to talk** mode (marked PTT), the **voice-activated microphone** (VOX) mode, or the **continuously open microphone** (HOT) mode. Five channels are available—two air-to-ground, one air-to-air, and two intercom. The switch positions are T/R (**transmit and receive**), RCV (**receive only**), and OFF. Two ATUs, one in the aft crew station and the other in the mid-deck, have a speaker-microphone built in so you can use them without the headset.

If you prefer not to have loose cables about, you can use the wireless microphone. It consists of a small transmitter which attaches to your headset.

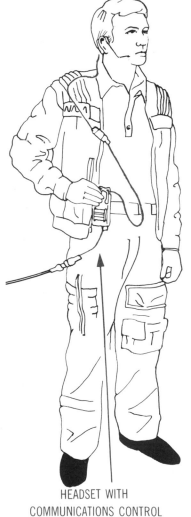

HEADSET WITH
COMMUNICATIONS CONTROL

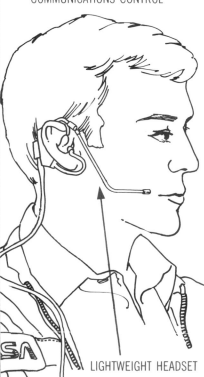

LIGHTWEIGHT HEADSET

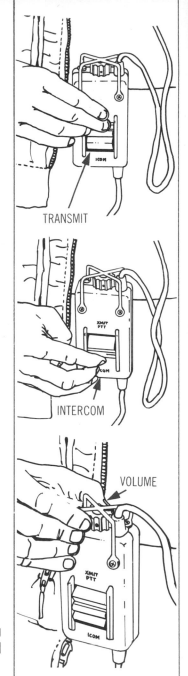

TRANSMIT

INTERCOM

VOLUME

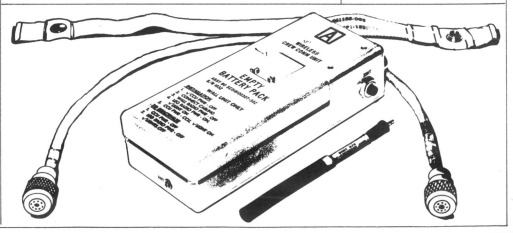

Mission controllers can transmit up to a hundred pages of information to you for storage in the computer. The data can then be recalled and displayed on a television screen. You can record the display with the Polaroid camera provided.

Before the Space Transportation System, communication with low-orbiting spacecraft was difficult. Information could be received only when a spacecraft was within range of a ground station, usually for only 5 minutes at a time per station. The oceans and geographical or political restrictions on locating ground stations meant that many satellites were out of touch most of the time.

Two satellites called Tracking and Data Relay Satellites (TDRSs) solve this problem. (See 3.27) They are in geostationary orbit, 130° apart. With the duo, it is possible to communicate with an orbiting spacecraft at least 85% of the time.

The TDRS system can handle up to forty satellites at a time. A nearly identical satellite, called Advanced WESTAR, is used for domestic communications. Another vehicle orbits as a spare. Your transmissions are beamed to either TDRS East or West (whichever is closer) and are then rebroadcast to a single ground station at White Sands, New Mexico.

TRACKING AND DATA RELAY SATELLITE SYSTEM

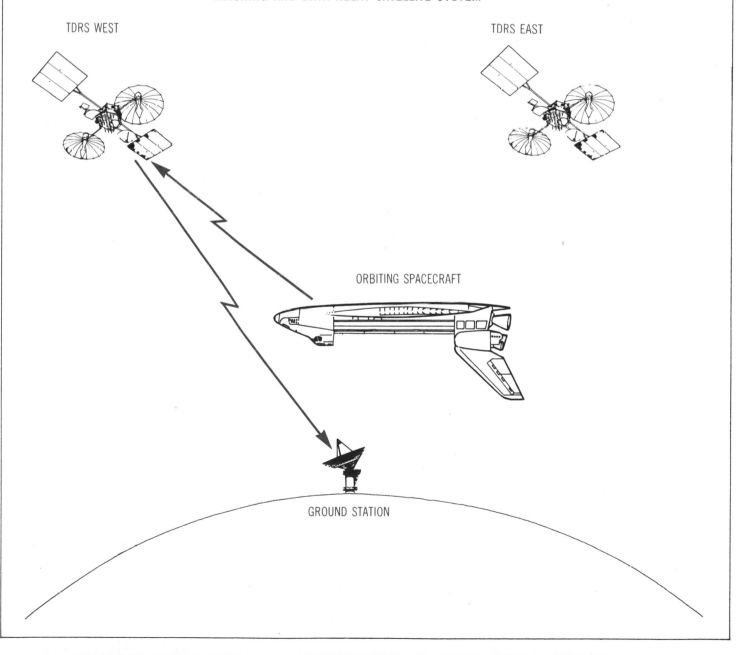

TDRS WEST

TDRS EAST

ORBITING SPACECRAFT

GROUND STATION

The Shuttle is a workhorse, carrying many different **payloads** into space. Shuttle payloads are classed as either attached or free-flying. **Attached payloads** such as Spacelab remain in the cargo bay throughout the mission. **Free-flying payloads**, as their name implies, are released to fly alone. Some free-flyers such as Space Telescope and the Multimission Modular Spacecraft are meant to be serviced or retrieved by the Shuttle. Others are boosted into orbits beyond the Shuttle's reach.

The following sections provide information on the major payloads you will handle. Scientists and engineers depend on the mission and payload specialists to do their job in space and to return data and services for those on the ground. It is important to understand both the kind of payload on your flight and the procedures for operating that payload.

Spacelab is an orbiting laboratory built by the European Space Agency for use with the Shuttle. It is an attached payload; that is, it remains in the Orbiter's cargo bay throughout the flight. This versatile facility was designed to provide the scientific community with easy, economical access to space. Although it is a payload, Spacelab is part of the Space Transportation System.

Like many other STS components, Spacelab is constructed from self-contained segments or modules. Spacelab has two major subsections: cylindrical, pressurized **crew modules** and U-shaped unpressurized instrument-carrying **pallets**. A crew module provides a "shirtsleeve" environment where payload specialists work as they would in a laboratory on the ground; pallets accommodate experiments for direct exposure to space. Crew modules and pallets are combined in a number of configurations to suit the needs of a particular mission. Possible combinations include a crew module with one, two, or three pallets, up to five pallets without a crew module, or a crew module without pallets.

Because of Spacelab's versatility, it can support a variety of scientific missions, including investigations in materials processing, life sciences, space physics, and remote sensing of earth. The varied nature of Spacelab missions makes it necessary to describe the facility in general terms only in this manual. Spacelab equipment is highly adaptable to the particular requirements of specific missions.

An oxygen-nitrogen atmosphere identical to that in the Orbiter crew compartment is maintained in the crew module. Depending on mission requirements, crew modules consist of either one segment (**short module**) or two (**long module**). The short module is 14 feet (4.2 meters) long; the long module measures 23 feet (7 meters). All crew modules are 13.1 feet (4 meters) in diameter. Most of the equipment housed in a short module is for controlling pallet-mounted experiments. The long module is used whenever more room is required for laboratory-type investigations. Equipment inside the crew modules is mounted in 19-inch (50-centimeter)-wide racks. These racks are easily removed between flights so module-mounted experiments can be changed quickly.

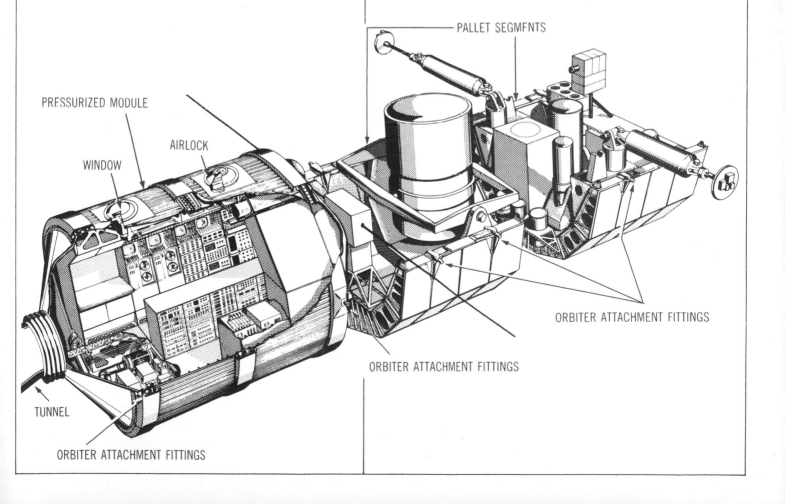

PALLET SEGMENTS

PRESSURIZED MODULE

AIRLOCK

WINDOW

ORBITER ATTACHMENT FITTINGS

TUNNEL

ORBITER ATTACHMENT FITTINGS

ORBITER ATTACHMENT FITTINGS

What sort of equipment can you find in the crew module? If you are on a life-sciences mission, there are animal cages and monkeys, rats, and mice. On a materials-processing flight, there will be a furnace and several heaters to create new alloys, crystals, and other materials. Such experiments investigate the use of the microgravity environment for industrial purposes. There are also data-recording devices and computers in the module.

Spacelab pallets are essentially open cargo containers. They serve as a stable platform for experimental equipment. Experiments mounted on a pallet are exposed to space. Telescopes, radiation counters, and earth-sensing equipment are types of pallet-borne hardware.

Some of the electronic controls for pallet-mounted experiments have to be shielded from the temperature extremes encountered in space. These controls are either contained in the crew module or, for pallet-only missions, are housed in a small cylindrical "igloo" attached to the pallet's end. Experiments on such pallet-only flights are controlled from the aft crew station. An individual pallet can hold up to 6,000 pounds of equipment.

A pressurized tunnel connects the crew module to the Orbiter mid-deck. Nobody is in Spacelab during launch—you don't crawl through the tunnel until you are safely in orbit. The airlock used for EVAs or spacewalks is attached to the tunnel. If you perform a spacewalk, use the airlock the same way you would if it were in the mid-deck. (Don't confuse the EVA airlock with the experiment airlock in the crew module. The latter is much smaller and is used only for experiments.)

Most of the non-astronaut payload specialists flown will be on Spacelab missions. Since Spacelab is an international effort, many of these will be European. If you are involved in one of these flights, you will have a rare opportunity to work in space alongside scientists and engineers from around the world.

SPACELAB WORK LOCATIONS

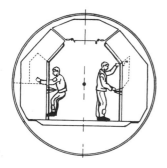

SOME SPACELAB CONFIGURATIONS

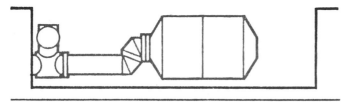

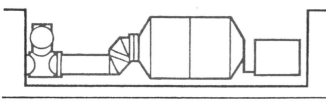

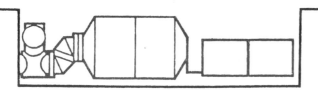

Space Telescope is one of the most important scientific payloads carried by the Shuttle. Astronomers place the age of the universe at between 10 and 18 billion years. Understanding the age and the amount of matter in the universe may answer not only the question of how it began, but what the ultimate fate of the universe will be. Will it collapse into a cosmic oblivion, or continue to expand forever until the stars burn out and the galaxies are just dying embers?

Earth-based astronomers are limited by living under an ocean of air that distorts the images reaching their telescopes. These earth-based telescopes can look out about 2 billion light-years. Since looking out is also looking back in time, the farther you can see, the more history you can record.

No matter how large the telescope or how clear the night, there is always distortion on earth. In space there is no atmosphere to hamper astronomical observations. From such a vantage point, a telescope would detect objects at incredible distances and make measurements impossible on earth. The Space Telescope will be able to see objects 50 times fainter than its earth-based counterparts. Faint galaxies detected will give astronomers a new image of the structure of the universe, and we may finally discover what powers the awesome and mysterious quasars.

With the ability to see 7 times farther—out to about 14 billion light-years—the Space Telescope may provide astronomers with a look at the "beginning," or at least indicate how far back on the cosmic timescale the universe goes.

We can only guess at the new objects—as surprising as quasars, pulsars, and black holes—the Space Telescope may uncover. Looking at familiar objects, the telescope will study the planets and will be used to try to detect planets around nearby stars. If planets are common, the chances are greater that other intelligent civilizations exist. Then radio telescopes could be used to detect any transmissions they make.

This orbiting observatory is a reflecting telescope with a 94-inch (24-meter)-diameter **primary mirror**. It is housed in a structure 43 feet (13 meters) long and 14 feet (4.2 meters) wide. Its total weight exceeds 13 tons (11,800 kilograms), well within the 32-ton (29,500-kilogram) capacity of the Shuttle. The telescope is an unmanned free-flying payload. After it is released, the data it collects is telemetered to astronomers on the ground. Target selection and pointing are controlled from the ground.

The 94-inch (2.4-meter) mirror is aligned from the ground. A smaller **secondary mirror** takes the light gathered by the primary mirror and directs it to one of the scientific detectors. Six electric motors precisely direct the smaller mirror. The detectors can see visible light as well as ultraviolet and infrared radiation blocked by the atmosphere. One detector will split the light into a spectrum to study chemical composition and radiation, and perhaps to locate black holes, which are themselves invisible, by analyzing their effects on companion stars. Other detectors will study nearby objects, the most distant objects, the variation in star motions, and the subtlest differences in brightness.

The Space Telescope is launched from the Kennedy Space Center. When the Shuttle is in orbit 500 miles (800 kilometers) above the earth, the telescope is deployed, using the remote manipulator arm. Following release, two large arrays of solar cells unfold and provide electrical power to the satellite.

Since the telescope is designed to last for 15 years, you may be called upon to change one or more of its scientific instruments. In an extreme case, you can even place the Space Telescope in the cargo bay for return to earth. It can then be repaired, overhauled, and returned to space.

Free-Flying Payload Deployment

1. **Check the remote manipulator system arm.**

2. **Turn payload on; check its systems.**

3. **Attach arm to grapple on payload.**

4. **Maneuver Orbiter into proper position.**

5. **Release payload capture latches in cargo bay.**

6. **Using arm, maneuver payload out of the cargo bay and into proper position.**

7. **Make sure payload is stable, not wobbling on end of manipulator.**

8. **Release payload and maneuver Orbiter away.**

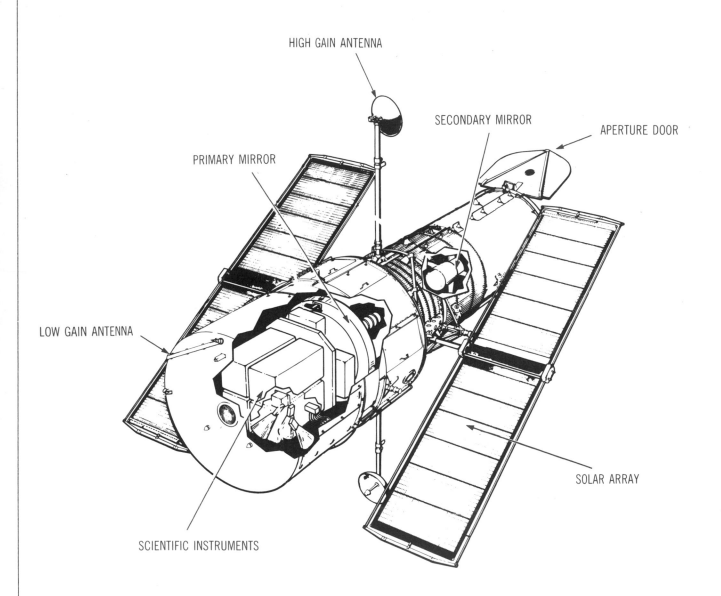

SPACE TELESCOPE

HIGH GAIN ANTENNA

SECONDARY MIRROR

APERTURE DOOR

PRIMARY MIRROR

LOW GAIN ANTENNA

SOLAR ARRAY

SCIENTIFIC INSTRUMENTS

The free-flying Long-Duration Exposure Facility, or LDEF (pronounced "el-def") for short, takes advantage of the Shuttle's ability to return objects from space. It is a cylindrical frame with shallow trays affixed to its exterior. Experiments that require long-term exposure to the space environment are housed in the trays. LDEF is carried aloft and released on one Shuttle flight, then retrieved on another mission several months later.

The 14-foot (4.2-meter)-diameter structure is made from aluminum I-beams. It is 30 feet (9 meters) long, and it holds seventy-two trays on the sides and two on each end.

Experiments may be either active or passive. **Passive experiments** usually consist of samples of such materials as mirrors, lenses, paints, and metals. Spacecraft designers are interested in finding out what happens to different materials when they're in space for a long time. **Active experiments** include any experiment that has a power source, data recorder, or other electronic compo-nents. These must be totally self-contained in their trays.

If you are on an LDEF mission, deploy it just as you would any other free-flying payload. Usually LDEFs are released in a 350-mile (560-kilometer)-high circular orbit. The payload has no reaction-control system. Instead, it uses something called **gravity gradient** stabilization. Within a week of being released, the payload will orient itself so that its long axis always points toward the earth.

The LDEF orbits alone for 6 to 9 months. Then it is retrieved and, using the manipulator arm, placed in the cargo bay. After the Orbiter lands and the LDEF is removed, the trays are returned to the experimenters, who can analyze the results at leisure.

If you perform an extravehicular activity while an LDEF is in the cargo bay, be very careful. Many of the experiments in the trays can be ruined by a careless touch of your hand or exhaust from your maneuvering unit.

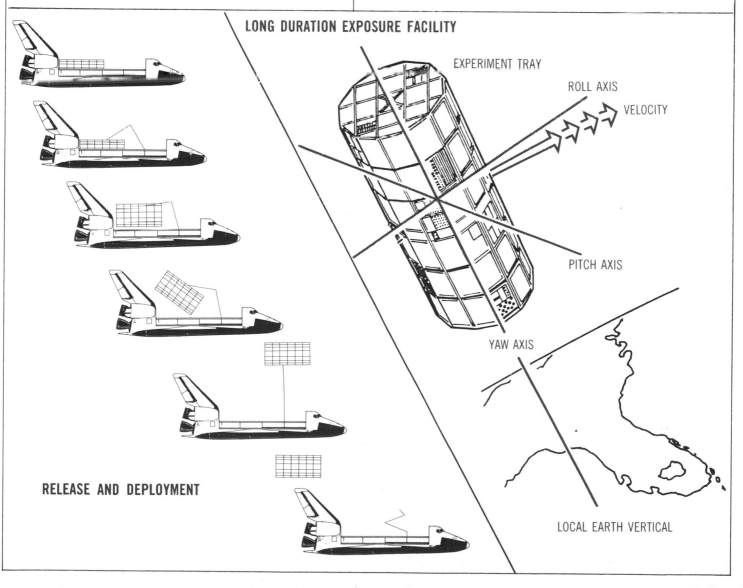

LONG DURATION EXPOSURE FACILITY

EXPERIMENT TRAY

ROLL AXIS

VELOCITY

PITCH AXIS

YAW AXIS

LOCAL EARTH VERTICAL

RELEASE AND DEPLOYMENT

Accessible to anyone with a good idea and $3,000, the **Small Self-Contained** (SSC) **Payloads** will probably be used to conduct experiments in several obvious and important areas. The detection of solar X-rays contributes to the understanding of the sun's internal processes. In electronics, efforts toward improving space-based computers and communication equipment could be tested. Processing of crystals, plastics, and metals in microgravity could bring important breakthroughs at very low cost. Finally, questions in biology requiring repeated, inexpensive access to space can be answered. One high school is launching an ant colony. Other experiments will look at the problem of bone decalcification and of changes in blood.

NASA's guidelines for these payloads place no more restrictions on them than necessary. First, they have to aid research or development. Also, they must weigh less than 200 pounds (91 kilograms), occupy less than 5 cubic feet (1.4 cubic meters), and require no Shuttle services. This dictates that they contain their own systems for power, handling data, and environmental control. You may have to turn them on or off from the aft crew station, but otherwise they are completely automatic. Finally, they must adhere to flight safety guidelines.

Housed in small cylindrical containers fastened to the forward end of the Orbiter's cargo bay, Getaway Specials are flown on a "space available" basis. That is, they will be placed aboard the Orbiter when allowed by space and weight restrictions. Chances are you will have one or more aboard on your flight—very few payloads take full advantage of the Shuttle's weight-lifting or volume-carrying capabilities.

Of course, you will be advised before launch whenever a Getaway Special payload is on board, but you won't need any payload-oriented training. Just turn the experiment on and off at the times indicated in your flight plan.

The Getaway Special concept has been very popular. By the time **Columbia** made her first flight in April 1981, more than three hundred had been sold.

SMALL SELF-CONTAINED PAYLOADS ARE ATTACHED TO THE INSIDE WALL OF THE CARGO BAY

Tracking and Data Relay Satellite

The Tracking and Data Relay Satellite (TDRS) is a space-based communications relay that links flight controllers on earth with orbiting spacecraft. Two such satellites are used, each in a geostationary orbit 22,300 miles (35,900 kilometers) above the earth. At that altitude it takes 24 hours to complete a single orbit. A satellite in a geostationary orbit over the equator remains fixed over the same point on the ground.

Two relay satellites spaced 130° apart (one over the Pacific Ocean, the other over the Atlantic) in orbit can communicate with a low-orbiting spacecraft during most of its flight. A small "zone of exclusion" exists where the lower vehicle is out of contact, but this lasts only a few minutes every orbit.

TDRS is a large satellite. When in position, it measures 57 feet (17 meters) long and 42.6 (13 meters) feet wide. For launch, it is folded to fit in the cargo bay. An inertial upper stage is attached to the satellite's base. After the satellite is released from its launch berth in the cargo bay, the upper-stage motor places it in geostationary orbit. When it is finally in position, the satellite unfolds and begins operating.

A total of four TDRS satellites are used. Two are dedicated to the TDRS system. One is used as a commercial communications satellite and is called the Advanced WESTAR. The fourth is a spare in case one of the other three is overloaded or breaks down.

Tracking and Data Relay Satellite Statistics

Size:	(deployed) 57 feet by 42.6 feet (17 meters by 13 meters)
Power:	(solar cells) at least 1850 watts
Service:	Up to twenty simultaneous satellites including the Shuttle
Antennas:	7
Orbit:	22,300 miles (35,900 kilometers), circular
Lifetime:	10 years
Channels:	2 S-band (duplex)
	2 K-band (duplex)
	1 twenty-user S-band multiple-access return
	1 time-shared S-band multiple-access forward

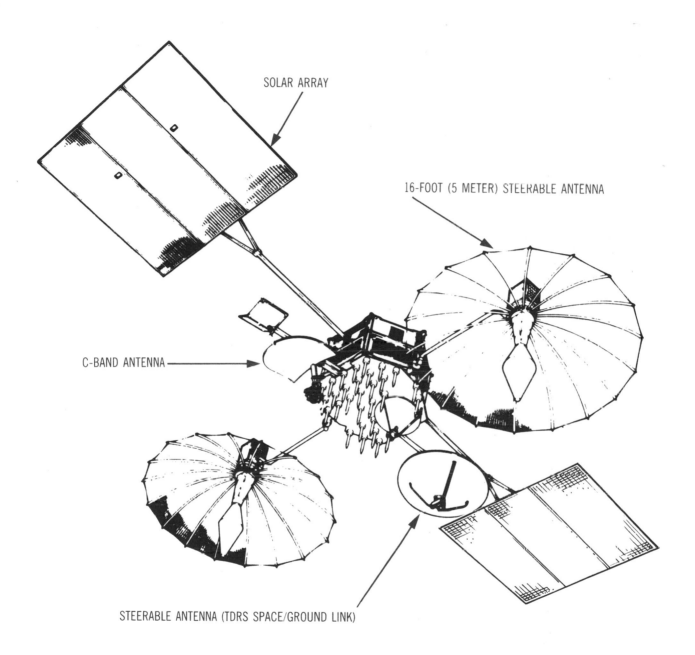

SOLAR ARRAY

16-FOOT (5 METER) STEERABLE ANTENNA

C-BAND ANTENNA

STEERABLE ANTENNA (TDRS SPACE/GROUND LINK)

Nearly all spacecraft have separate systems for power, data handling and communication, and attitude control. The Multimission Modular Spacecraft (MMS) represents an attempt to incorporate these systems in a standardized support vehicle that is attached to the payload. Hence, MMS-derived satellites comprise two distinct segments: the payload section (or instrument module) and the support section.

The MMS consists of a series of self-contained sections, or modules, which attach to the sides and bottom of a triangular truss; the instrument module attaches to its top. Each module houses one subsystem. For example, the attitude-control system is assembled into a plug-in unit. Modules for communication and data handling, and for power if needed, complete the MSS. Components for the modules (batteries, voltage regulators, tape recorders, etc.) are drawn "off the shelf," allowing a spacecraft's capabilities to be tailored to specific mission requirements. This standardization of parts and the "building block" approach to spacecraft design reduces both cost and fabrication time and makes it easier for you to repair these satellites in orbit.

MMS satellites are launched by both the Space Shuttle and by expendable launch vehicles. The first such spacecraft flown was the Solar Maximum Mission, launched on February 14, 1980, by a Delta. This satellite had a remote-manipulator grapple on it to facilitate later retrieval by the Shuttle. The latest Landsat also uses the MMS. MMS satellites can be returned to earth and their parts reused.

An assembly called the **flight support system** (FSS) fits in the Orbiter cargo bay and holds the MMS during launch and landing. The FSS is also used for deploying, retrieving, and servicing satellites in orbit. The FSS consists of a spacecraft-retention cradle, a payload-positioning platform, a module-exchange mechanism, and a module magazine. The horseshoe-shaped **cradle** is the connection between the cargo bay and MMS instrument module; it fits in the cargo bay and the instrument module rests inside it. The rotating positioning **platform** erects the MMS for deployment. The module magazine and exchange mechanism are required on servicing missions. The **magazine** holds replacement modules; the **exchange mechanism**, controlled from the aft crew station, removes an old module from the MMS and replaces it with a new one from the magazine.

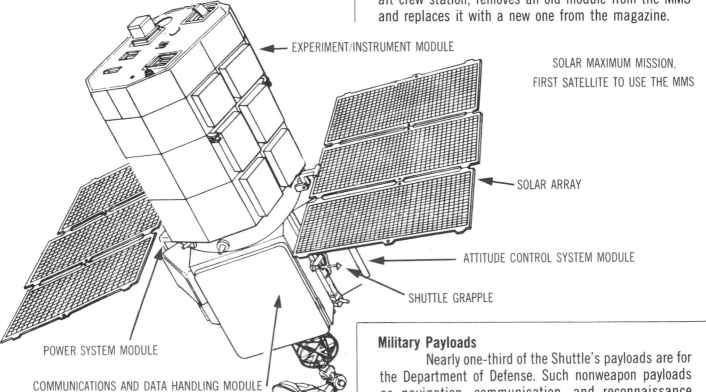

EXPERIMENT/INSTRUMENT MODULE

SOLAR MAXIMUM MISSION, FIRST SATELLITE TO USE THE MMS

SOLAR ARRAY

ATTITUDE CONTROL SYSTEM MODULE

SHUTTLE GRAPPLE

POWER SYSTEM MODULE

COMMUNICATIONS AND DATA HANDLING MODULE

HIGH GAIN ANTENNA

Military Payloads

Nearly one-third of the Shuttle's payloads are for the Department of Defense. Such nonweapon payloads as navigation, communication, and reconnaissance satellites are carried. These payloads are usually classified, and the Defense Department supplies its own crews for these missions. However, you may have to perform emergency maintenance on one of these satellites. If so, Mission Control will transmit special instructions to you.

On every flight, there is a chance
that something will go wrong.
This section describes possible
malfunctions and what you'll have to do
in the resulting emergency.

Launch aborts fall into two categories, those that occur on the launch pad and those that happen during ascent. An **abort** is necessitated by an emergency that must be corrected immediately and prevents you from completing the planned mission.

Launch-pad aborts require you to evacuate the Orbiter quickly. If the access arm with the **white room** has already retracted, it will move into position so you can exit through the hatch. Five **slide wires** extend from the service structure to the entrance of an underground bunker 1,200 feet (375 meters) away. There's a steel basket on each wire. Each of the five baskets can hold two people. Before you release the brake on your basket and slide down the wire, make sure everybody is in a basket. When all your crewmates are ready, release the baskets. They'll slide down the wires to the landing zone in about 35 seconds. After you've stopped, get out of the basket and go to the bunker. A pad abort is possible anytime before the solid boosters ignite. If one of your main engines malfunctions during thrust build-up, all three shut down and the launch is canceled.

EMERGENCY EXIT SYSTEM

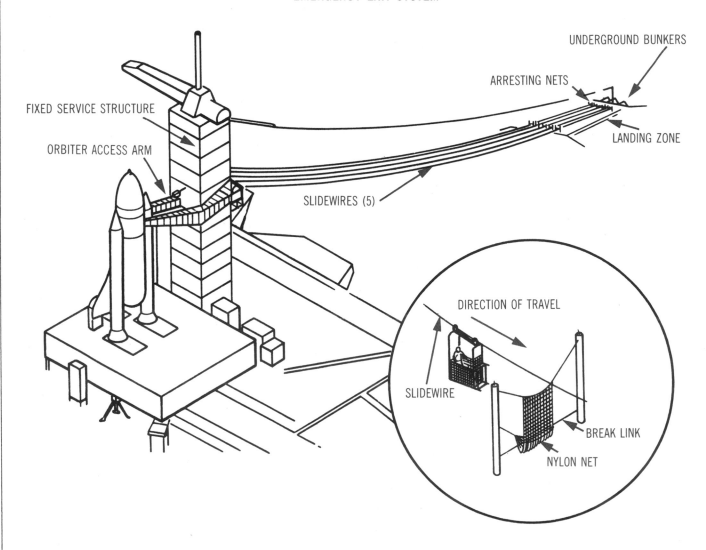

UNDERGROUND BUNKERS

ARRESTING NETS

FIXED SERVICE STRUCTURE

ORBITER ACCESS ARM

LANDING ZONE

SLIDEWIRES (5)

DIRECTION OF TRAVEL

SLIDEWIRE

BREAK LINK

NYLON NET

Once the Solid Rocket Boosters ignite, though, you are committed to at least a partial flight. A launch abort occurs if one or two of your Orbiter's main engines fail. Depending on when the failure occurs, several abort options exist. In order, they are: **return to launch site** (RTLS); an **emergency landing** at the U.S. Naval Air Station in Rota, Spain; **abort once around** (AOA); and **abort to orbit** (ATO).

The RTLS option is used if the main propulsion system malfunctions in the first 4 minutes and 20 seconds of flight. An RTLS does not begin until after the solid-propellant boosters have burned out and have been jettisoned. The Orbiter and External Tank continue downrange on the power of the remaining main engines, both OMS engines, and the four aft-firing maneuvering rockets. You'll continue until just enough propellants remain in the tank for you to reverse your flight direction. A 5°-per-second pitch begins until you're pointing back toward the launch site. When your propellants are exhausted, the main engines cut off and the tank is discarded. You'll then pilot the Orbiter back to the launch site and land as you would after a normal flight.

The next abort option—an emergency landing at the Naval Air Station in Spain—applies only to launches from the Kennedy Space Center. It is used if a main engine fails and you are unable either to turn around or to go on for an abort once around.

The AOA mode is used when one or two main engines fail after the solid boosters burn out and before an abort to orbit is possible. In an AOA, continue firing the remaining main engines and, as with the RTLS, use the OMS and maneuvering engines. Continue to fire the OMS and RCS engines until there's only enough propellant left to fire the OMS engines twice after main engine cut-off. When you've exhausted the propellants for the main engines, jettison the External Tank just as you would in a normal flight. Now fire the OMS engines to give yourself the altitude and speed you'll need to reach the emergency landing strip in White Sands, New Mexico. Since White Sands is four-fifths of the way around the world from your launch site, another OMS firing will be needed to sustain your suborbital flight. Finally, enter the atmosphere and land at White Sands just as you would at the primary landing site.

The last launch abort is an abort to orbit. The ATO is used if a main engine fails late in the ascent. As with other abort profiles, keep firing the remaining main engines. Slightly longer-than-usual OMS burns will enable you to reach a lower orbit than planned. You'll have to go to an alternate mission plan, but you'll probably be able to complete most of your mission objectives. Deorbit, entry, and landing are similar to those of a normal mission.

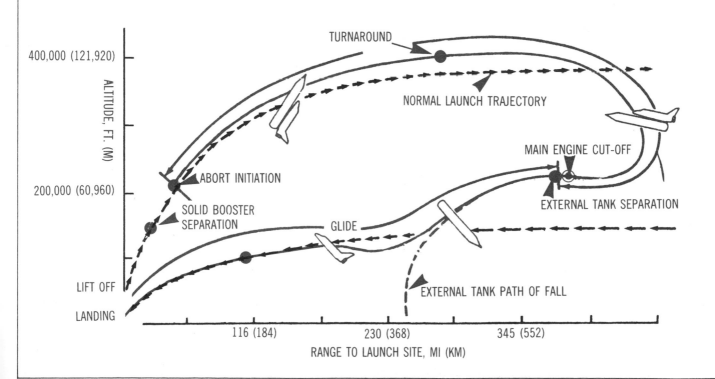

RETURN TO LAUNCH SITE ABORT

TURNAROUND

400,000 (121,920)

NORMAL LAUNCH TRAJECTORY

MAIN ENGINE CUT-OFF

ALTITUDE, FT. (M)

ABORT INITIATION

200,000 (60,960)

SOLID BOOSTER SEPARATION

EXTERNAL TANK SEPARATION

GLIDE

EXTERNAL TANK PATH OF FALL

LIFT OFF

LANDING

116 (184)　　230 (368)　　345 (552)

RANGE TO LAUNCH SITE, MI (KM)

An equipment malfunction in space can be very serious. If a major system fails, it will at least mean you'll have to end your flight early. It could also require a rescue mission by another Orbiter.

One in-flight emergency that can occur during ascent and is relatively easy to remedy is having the Solid Rocket Boosters fail to separate automatically. If this should happen, flip the SRB SEP mode switch on panel C3 to the MAN/AUTO position. Lift the cover over the SEP (separation) button and push the button. This jettisons the motors. Another relatively easy-to-correct problem is failure of the main engines to throttle automatically. All you have to do is use the throttle/speedbrake handle on your left. Move the handle to control main-engine thrust.

Other, graver problems can occur in orbit. Although the Orbiter's electronics are air-cooled, it's possible that a part will malfunction, overheat, and catch fire. For onboard fires, there are three built-in remote-control fire extinguishers, and four portable ones. **Smoke detectors** are located in all three avionics (electronic equipment) areas in the mid-deck and the inhabited areas of the crew compartment. Controls for the detectors and the built-in fire extinguishers are on panel L1 on the flight deck. If a fire or smoke is detected, the appropriate panel light illuminates, showing the fire's location, and an alarm sounds. Upon hearing the alarm, each crew member must don a portable oxygen system (POS). If the fire is in one of the avionics bays, the AV BAY 1, 2 or 3 light will be on. Arm the FIRE SUPPRESSION system for that bay, and push the AGENT DISCH button to actuate the fire extinguisher.

Use the portable fire extinguishers for fires in other cabin areas. Each instrument panel has a FIRE HOLE. To combat a fire behind a panel, put the extinguisher's nozzle into the port and empty the extinguisher.

Still another class of emergencies involve cabin atmosphere. If the problem is with the oxygen life-support system, use the emergency supply. This is controlled from panel L2. First, close the O_2 SYS 1 and SYS 2 valves. This allows the cabin pressure to decrease to 8.0 psi (55 kilopascals). Next, turn on the emergency oxygen supply: O_2 EMER to OPEN. At this time, don your POS. Use the hose and Orbiter oxygen to conserve the POS pack's own supply. If you can't fix the problem, a rescue mission will be necessary.

FIRE EXTINGUISHER SYSTEM

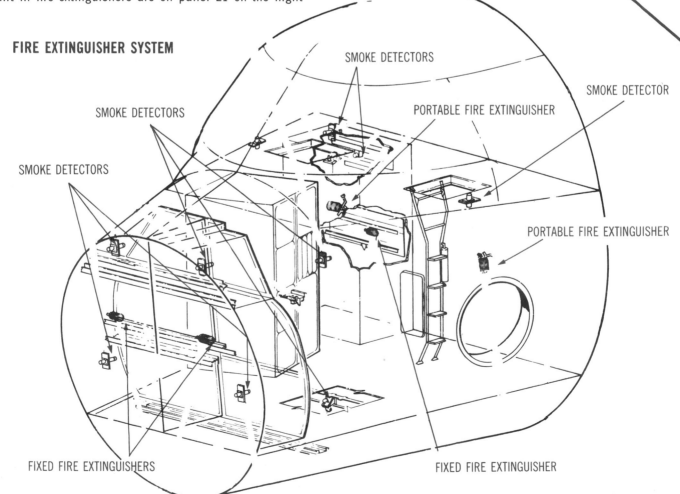

SMOKE DETECTORS

SMOKE DETECTORS

SMOKE DETECTORS

SMOKE DETECTORS

SMOKE DETECTOR

PORTABLE FIRE EXTINGUISHER

PORTABLE FIRE EXTINGUISHER

FIXED FIRE EXTINGUISHERS

FIXED FIRE EXTINGUISHER

FIRE EXTINGUISHER CONTROL PANEL

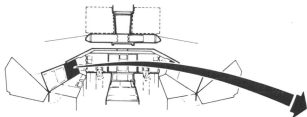

NOTE

(1) TO DISCHARGE, PULL AND POSITION FIRE SUPPRESSION SWITCHES TO ARM POSITION.

(2) LIFT GUARD AND DEPRESS AGENT DISCH PUSHBUTTON FOR ONE SECOND. (PUSHBUTTON LIGHTS UPON DISCHARGE.)

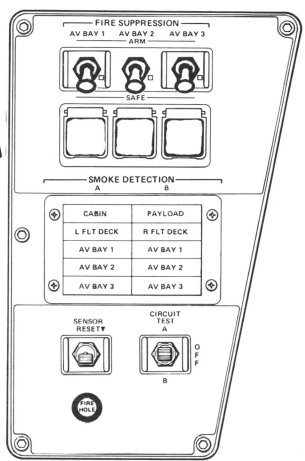

CABIN	PAYLOAD
L FLT DECK	R FLT DECK
AV BAY 1	AV BAY 1
AV BAY 2	AV BAY 2
AV BAY 3	AV BAY 3

You and your crew will have to transfer to the rescue craft. Since there are only two spacesuits on board, **personal rescue enclosures** are provided for the rest of the crew. The enclosure, or **rescue ball**, is a 30-inch (76-centimeter)-diameter fabric sphere. The nonspacesuited crew member opens the ball by its zipper. Wearing the POS, he gets into the ball; another member of the crew zips it shut. This way, the spacesuited astronaut can transfer him to the rescue ship during a spacewalk. The astronaut should connect the ball to the Orbiter's oxygen system until just before the transfer —this conserves the POS pack's internal oxygen until it absolutely must be used.

Another scenario for using the rescue ball is enacted if a toxic gas contaminates the cabin's air. After everybody is in a spacesuit or rescue ball, the astronauts in the suits depressurize the whole cabin. This vents the air, including the poison gas. After the cabin is repressurized, everybody doffs their rescue ball or spacesuit.

PORTABLE OXYGEN SYSTEM

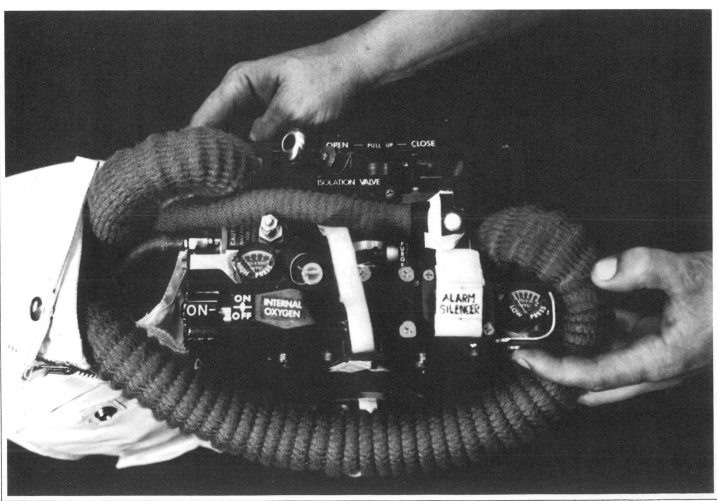

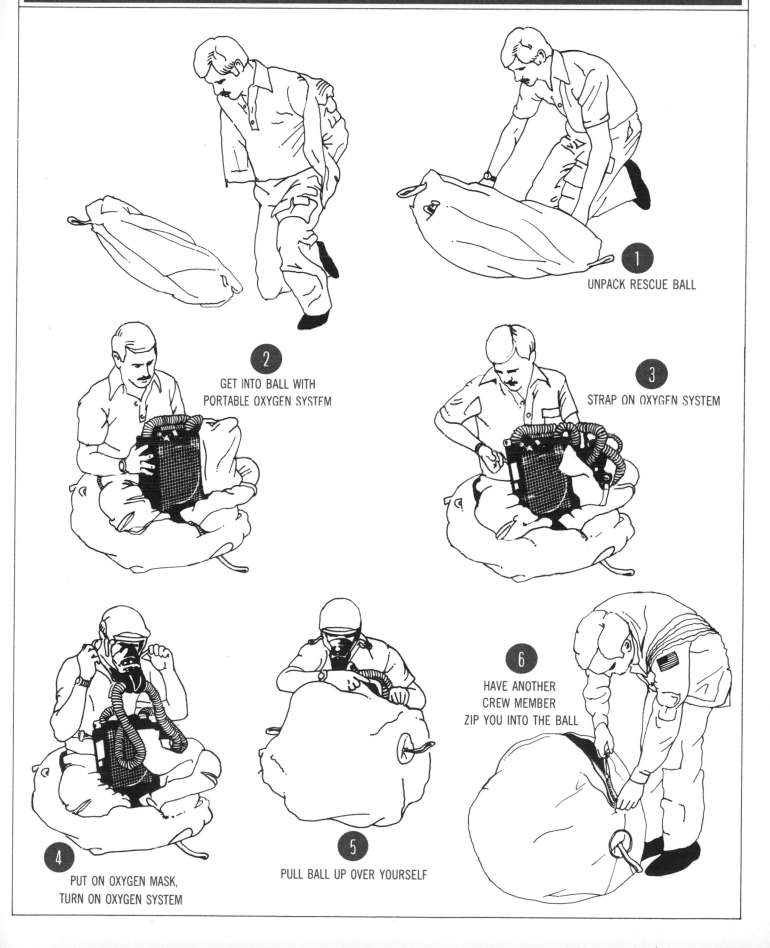

1 UNPACK RESCUE BALL

2 GET INTO BALL WITH PORTABLE OXYGEN SYSTEM

3 STRAP ON OXYGEN SYSTEM

4 PUT ON OXYGEN MASK, TURN ON OXYGEN SYSTEM

5 PULL BALL UP OVER YOURSELF

6 HAVE ANOTHER CREW MEMBER ZIP YOU INTO THE BALL

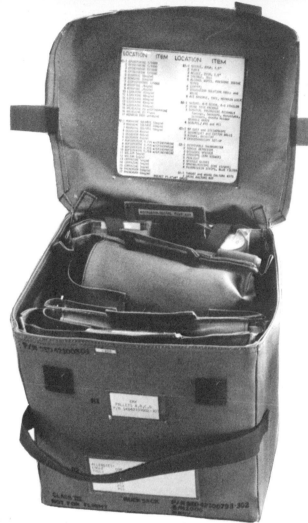

The Shuttle's **medical kit** allows you to care for any simple illnesses and injuries that occur during your flight. The kit also has equipment that allows you to stabilize severely ill or injured crew members until you can return to earth.

The kit is contained in three fabric packages. There's an emergency medical pack, a bandages-and-medications pack, and a pack with a defibrillator, an intravenous fluid system, and a respirator. Together, the three packs weigh less than 18 pounds (8 kilograms).

Such items as a stethoscope, blood-pressure cuff, sutures, disposable thermometers, and injectable medications are in the emergency pack. You'll find Band-Aids, adhesive tape, gauze bandages, and oral medicines in the bandages-and-medications kit.

The **Shuttle Orbiter medical system**, or SOMS, as the three-part medical kit is called, is stored in the mid-deck.

◄ MEDICAL KIT, STOWED

BANDAGES KIT

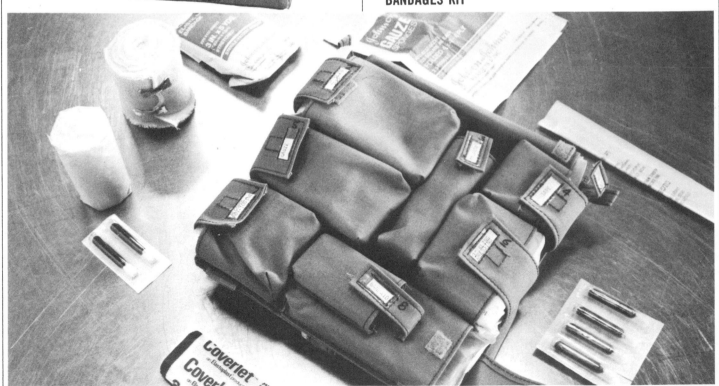

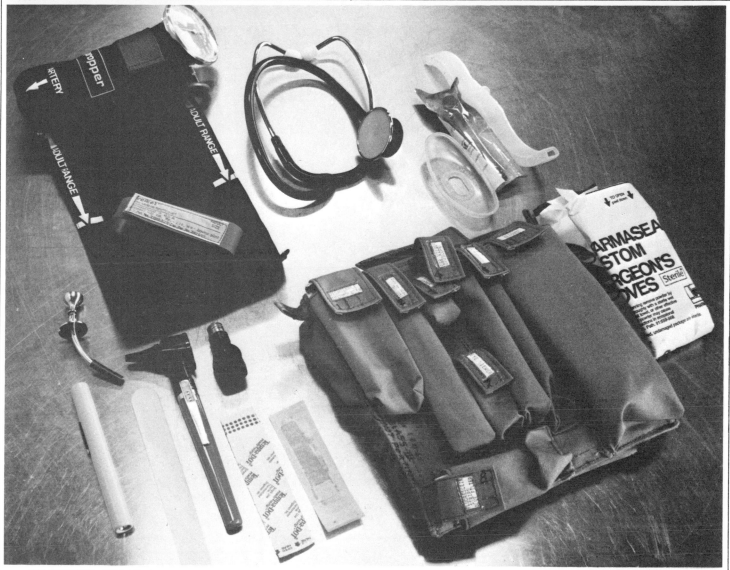

DIAGNOSTIC AND EMERGENCY KIT

ORAL MEDICATIONS

OINTMENTS, BANDAGES, AND NASAL SPRAY

Many science-fiction movies and novels have depicted astronauts stranded in orbit after their retro-rockets failed. Such a situation is **extremely** unlikely with the Shuttle. Normally, both **orbital maneuvering system** (OMS) engines fire during your deorbit maneuver. If one fails to fire, the other can slow you enough for your return. In the rather unlikely event that **both** OMS engines fail, you can use the maneuvering rockets. You'll need a deorbit burn that is 3½ times longer than a normal two-engine OMS burn, but it will work.

If you have to make an abnormal landing (caused, for example, by a landing-gear malfunction), there are a few special rules you'll have to follow. First, when the Orbiter comes to a stop, make sure everyone on board is all right. Report your crew's status to the recovery crew. For an emergency shutdown of the Orbiter after such a landing, follow this procedure using the switches on panel R1:

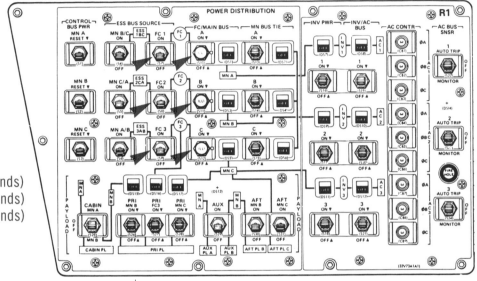

POWER DISTRIBUTION SWITCHES

FC/MAIN BUS A to OFF (hold for 2 seconds)
FC/MAIN BUS B to OFF (hold for 2 seconds)
FC/MAIN BUS C to OFF (hold for 2 seconds)
ESS BUS SOURCE, FC 1 to OFF
ESS BUS SOURCE, FC 2 to OFF
ESS BUS SOURCE, FC 3 to OFF

Once you've completed the emergency shutdown, leave the Orbiter. Use the crew hatch in the mid-deck whenever possible. There's a movable bar on the hatch.

When the hatch is open, swing the bar out. This gives you a handhold for jumping off the hatch. If you can't get the crew hatch open, the left-hand overhead window near the aft crew station is removable. You have a rope and a **descent device** that is used to climb down from the top of the Orbiter. The descent device controls your rate of descent down the rope. Whichever route you use to leave the Orbiter, make sure you drape the thermal curtains over the vehicle. The Orbiter's exterior is very hot (several hundred degrees) just after landing.

In case you have to land in a remote area, there's a survival kit onboard. It contains enough equipment to sustain a seven-person crew for 48 hours.

BAR

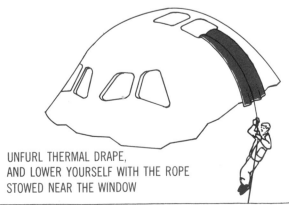

UNFURL THERMAL DRAPE,
AND LOWER YOURSELF WITH THE ROPE
STOWED NEAR THE WINDOW

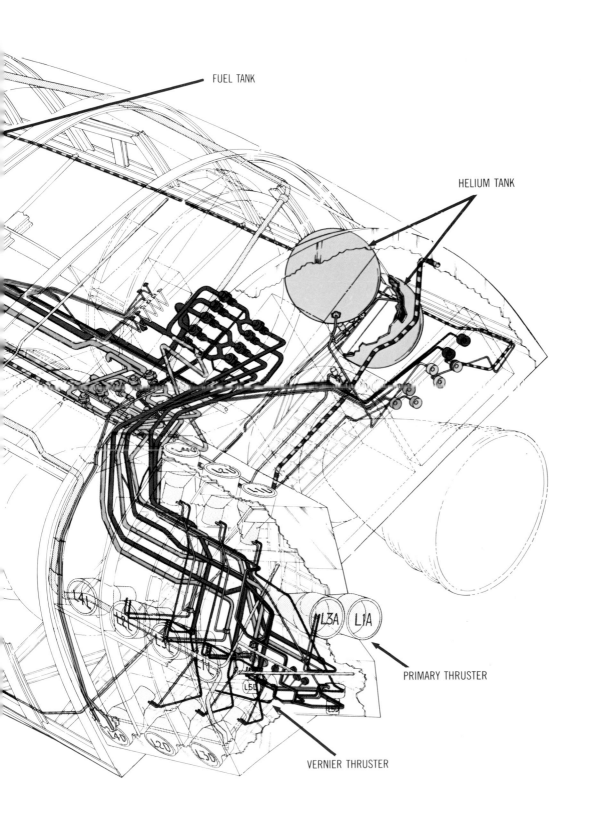

FUEL TANK

HELIUM TANK

L3A L1A

PRIMARY THRUSTER

VERNIER THRUSTER

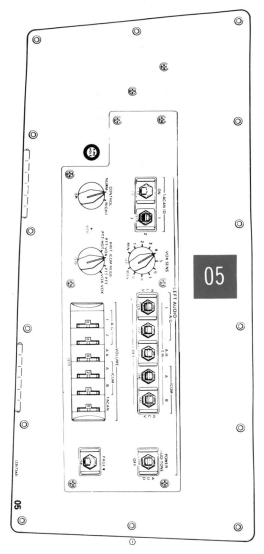

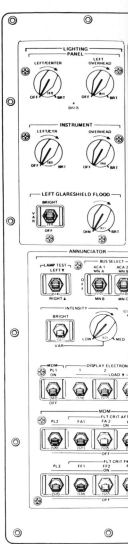

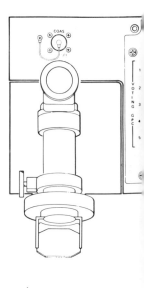

05

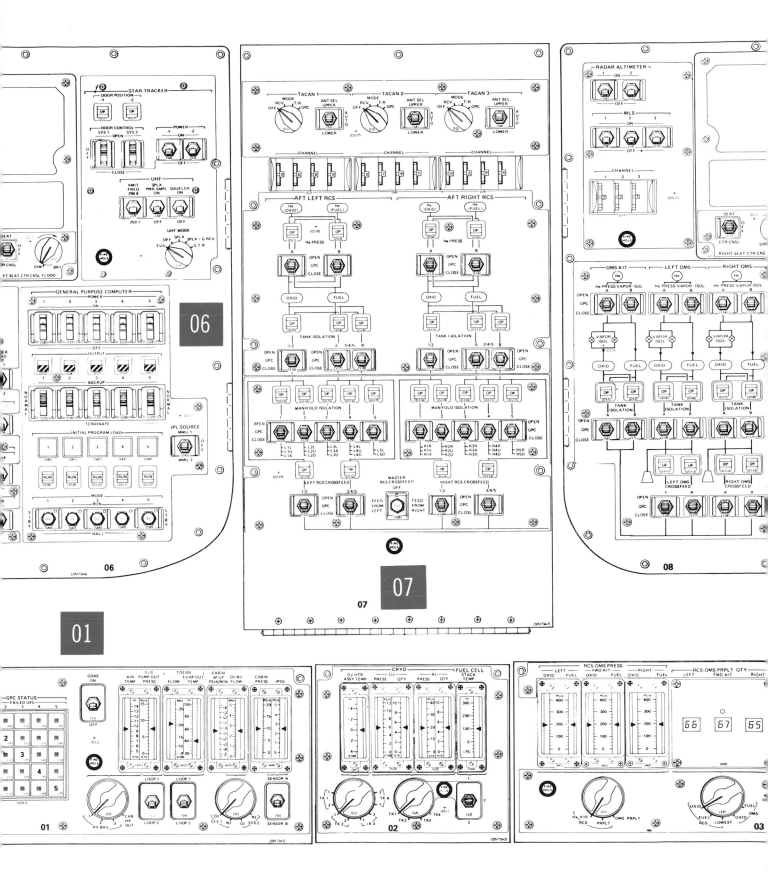

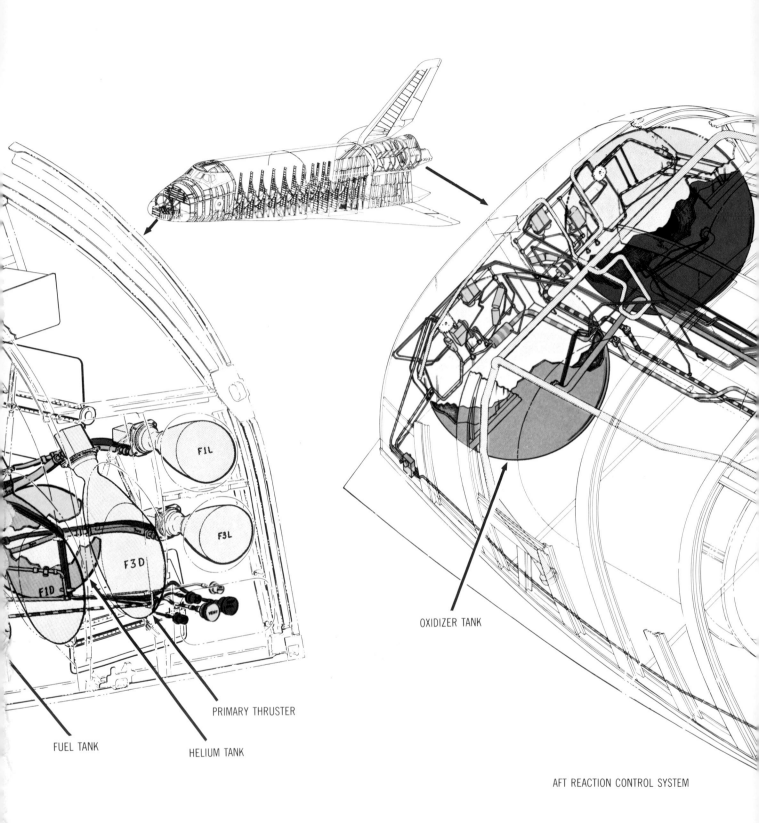

F1L

F3L

F3D

F1D

FUEL TANK

HELIUM TANK

PRIMARY THRUSTER

OXIDIZER TANK

AFT REACTION CONTROL SYSTEM

REACTION CONTROL SYSTEM

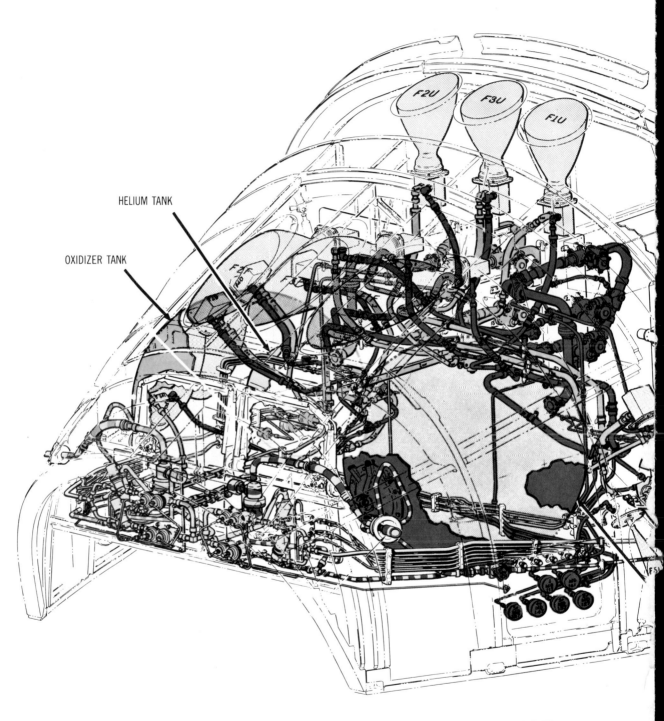

HELIUM TANK

OXIDIZER TANK

F2U F3U F1U

FORWARD REACTION CONTROL SYSTEM

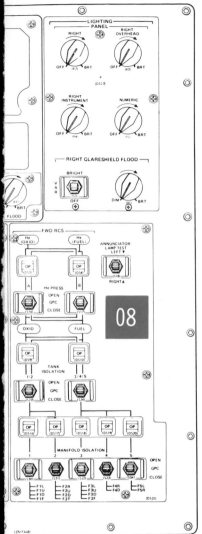

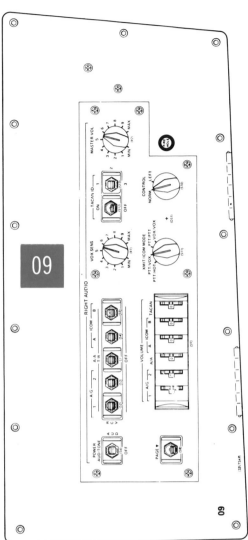

FLIGHT DECK CONSOLES
01, 02, 03, 05, 06, 07, 08, and 09

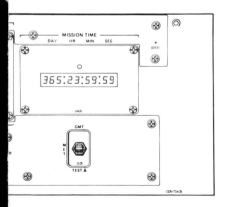

This section provides information for leaving orbit, entering the Earth's atmosphere and landing safely.

READ BEFORE LEAVING ORBIT

Deorbit, atmospheric entry, and landing are the final, most critical phases of your flight. You begin this part of your mission still orbiting 200 miles (320 kilometers) above the ground at a speed of over 17,000 mph (27,000 kph) and end it stopped on a runway half a world away. There are five major steps in the descent process: deorbit burn, entry, terminal-area energy management (TAEM), autoland, and touchdown.

To prepare for the **deorbit burn**, first turn the Orbiter around so that it is traveling tail-first. Fire the orbital maneuvering system (OMS) engines for 2 to 3 minutes. The exact firing time depends upon the weight of the Orbiter and its payload; you want to reduce your speed by about 200 mph (320 kph). After the OMS burn is over, turn the Orbiter around so it is traveling nose-first again. Raise the nose so it is pointing up at an angle between 28° and 38°. You began to lose altitude once you slowed down, so about half an hour after your deorbit burn, you will have descended to an altitude of 400,000 feet (122,000 meters).

This point in your flight is called **entry interface** and is where atmospheric entry officially begins. As the Orbiter descends, atmospheric drag dissipates its tremendous energy, which generates a great deal of heat. Following entry interface, the heat quickly builds up, and portions of the vehicle's exterior reach 1,540°C (2,800°F). The heat strips electrons from the air around the Orbiter, enveloping it in a sheath of ionized air that blocks all communication with the ground. The communications blackout lasts about 12 minutes.

During this interval, the Orbiter's reaction-control system begins to shut down and the aero-control surfaces take effect. You will also perform several banking maneuvers called **roll reversals** or **S-turns** to control your descent.

When you come out of the communications blackout, about 12 minutes before touchdown, you'll be committed to a particular landing site. You must begin your final approach with enough altitude and speed to reach the touchdown point. Follow the descent trajectory shown on the **computer display screens** (CRTs) very closely. This process of conserving your energy is called **terminal-area energy management**, or TAEM.

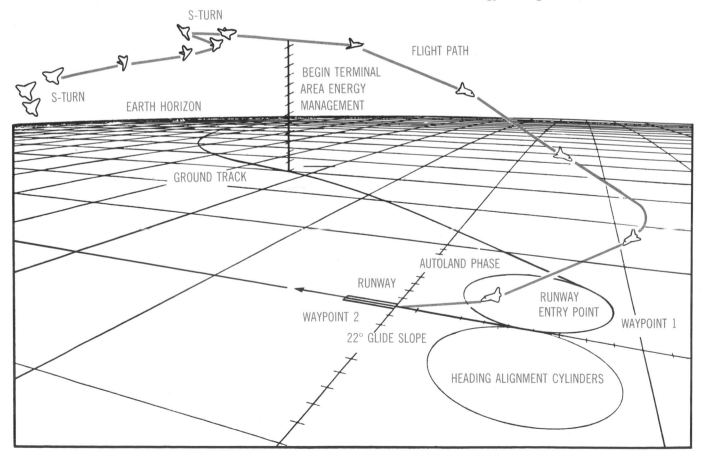

ENTRY FLIGHT PROFILE

Imagine that there are two 18,000-foot (5,500-meter)-diameter cylinders side-by-side about 7 miles (11 kilometers) away from the runway. During TAEM, you'll steer toward one of these cylinders, follow its curve, and line up with the runway. The point where you arrive at the cylinder is called **waypoint one**. You leave the cylinder at the so-called **runway entry point**.

Now you begin the **autoland phase**. You're descending on a 22° glideslope, straight toward the runway. Pull the nose up in a **flare** maneuver to reduce your glide angle to 1.5°, deploy the landing gear when you're only 90 feet (27 meters) off the ground (14 seconds before touchdown), and land. Throughout the autoland phase, a microwave landing system, the **tactical air navigation** (TACAN) locating system, and a **radar altimiter** provide landing data.

For your return from space, you'll use several instruments and controls, some of which you've used before for different functions. In the atmosphere, the rotational hand controller controls the elevons for pitch and roll like the stick in a conventional aircraft. Pedals operate the rudder for yaw, again like an airplane. When you're rolling down the runway, push down on both pedals simultaneously to operate the wheel brakes. Aerodynamic braking is provided by a speed brake controlled by the handle that was used to throttle the main engines during ascent.

All your flight instruments are on the front control panel. The computer display screens, or CRTs, show your position relative to the desired flight path. Lines on the display indicate the ideal trajectory or path and your upper and lower safe limits. A tiny Orbiter symbol represents your actual position. On the vertical-position display, a small box shows where you'll be in five minutes if you continue on your present course. Dots are used on the horizontal display to show your future location.

There are two circular instruments directly in front of you. The top one is the **attitude direction indicator** (ADI), also called the "8-ball." It shows pitch and roll; that is, it indicates what angle up or down the nose is pointing and shows whether or not the wings are level. The lower instrument looks like a compass and is called the **horizontal situation indicator** (HSI). It shows your location relative to various navigation points during entry, TAEM, and final approach. Your heading, the runway heading, glideslope deviation, and distance to the next navigation point are shown on the HSI.

On each side of the ADI and HSI, you'll find the **Alpha/Mach meter** (on the left) and the **altitude/vertical velocity indicator**.

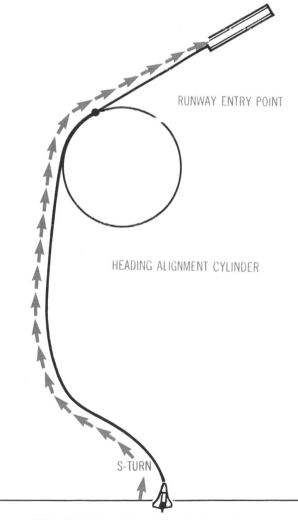

RUNWAY ENTRY POINT

HEADING ALIGNMENT CYLINDER

S-TURN

BEGIN TERMINAL AREA ENERGY MANAGEMENT

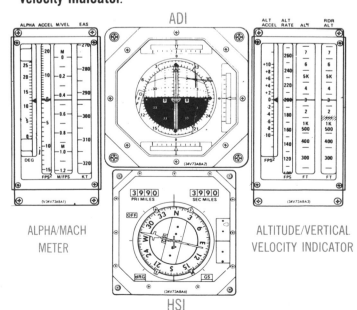

ALPHA/MACH METER

ADI

ALTITUDE/VERTICAL VELOCITY INDICATOR

HSI

DEORBIT BURN
60 MIN TO TOUCHDOWN
175 MILES (282 km)
16 465 MPH (26 498 km/H)

BLACKOUT
25 MIN TO TOUCHDOWN
50 MILES (80.5 km)
16 700 MPH (26 876 km/H)

MAXIMUM HEATING
20 MIN TO TOUCHDOWN
43.5 MILES (70 km)
15 045 MPH (24 200 km/H)

EXIT BLACKOUT
12 MIN TO TOUCHDOWN
34 MILES (55 km)
8275 MPH (13 317 km/H)

**TERMINAL AREA
ENERGY MANAGEMENT**
5.5 MIN TO TOUCHDOWN
83 130 FT (25 338 m)
1700 MPH (2735 km/H)

AUTOLAND
86 SEC TO TOUCHDOWN
13 365 FT (4074 m)
424 MPH (682 km/H)

12 695 MILES (20 865 km)	3392 MILES (5459 km)	1775 MILES (2856 km)	550 MILES (885 km)	60 MILES (96 km)	7.5 MILES (12 km)

AUTOLAND INTERFACE
86 SEC TO TOUCHDOWN
7.5 MILES (12 km) TO RUNWAY
424 MPH (682 km/H)
13 365-FT (4074-m) ALTITUDE

INITIATE PREFLARE
32 SEC TO TOUCHDOWN
2 MILES (3.2 km) TO RUNWAY
358 MPH (576 km/H)
1725-FT (526-m) ALTITUDE   22° GLIDE SLOPE

COMPLETE PREFLARE
17 SEC TO TOUCHDOWN
2540 FT (1079 m) TO RUNWAY
308 MPH (496 km/H)
135-FT (41-m) ALTITUDE   FLARE TO 1.5°

WHEELS DOWN
14 SEC TO TOUCHDOWN
1 100 FT (335 km) TO RUNWAY
268 MPH (430 km/H)
90-FT (27-m) ALTITUDE   1.5° GLIDE SLOPE

TOUCHDOWN
2760 FEET (689 m)
FROM END OF RUNWAY
215 MPH (346 km/H)

The vertical angle between the Orbiter's wings and its flight direction is called the **angle of attack**, referred to by the first letter of the Greek alphabet, **alpha**. There are four vertical scales on the Alpha/Mach meter. Your angle of attack is shown on the left scale. The next scale shows your acceleration in feet per second per second. The third scale indicates your speed relative to the speed of sound—the Mach number (Mach 1 is the speed of sound, Mach 2 is twice the speed of sound, etc.). Your airspeed in **knots equivalent airspeed** (KEAS) is displayed on the last scale.

The AVVI shows your radar and indicated altitudes and the rate in feet per second that your altitude is changing. Horizontal acceleration, in feet per second per second, is also shown.

On panel F7, between the upper CRTs, the **surface position indicator** shows the position of the aerodynamic control surfaces. Next to the bottom CRT, an **accelerometer** shows the g-forces during flight. During entry, you shouldn't experience more than 1.5 g.

A glass panel is mounted on the instrument panel glareshield, in front of the window. You look through the panel, which is called a **head-up display**, out the window. Your speed and altitude are projected on this display. This lets you look out the window and keep informed of your flight status at the same time.

Descent Checklist

Your deorbit, entry, and landing checklist has been compiled from the cue cards in your flight-data file. It has been simplified as a quick reference for your return. Events marked with an asterisk (*) occur automatically or require no action from you.

The times, speeds, and altitudes are approximate—changes in the Orbiter's weight and changing weather conditions mean the listed values vary from flight to flight. The dialog is a guideline for what you should say and hear. However, some transmissions must be made **precisely** as they're given here. Such dialog is enclosed in quotation marks.

Before you and the rest of the crew take your places, make sure all equipment has been stowed.

As a reminder, controls and displays are located as follows:

Panel	Location
F	Front
L	Left side, next to commander
C	Center console
R	Right side, next to pilot
O	Overhead
CRT	Cathode-ray tube (TV screen)

TIME	ALTITUDE/ VELOCITY	EVENT	PANEL	PROCEDURE	VOICE 5.6
(Landing minus hr: min:sec)					(Y = You; G = Ground Controller)
L–2:00:00	Orbit 17,300 mph (27,700 kph)	Return to seats.		Unstow seats for mission and payload specialists. All crew members to landing positions.	
L–1:40:00	Orbit 17,300 mph (27,700 kph)	Prepare for deorbit burn.	08	Check OMS engine status: He PRESS/VAPOR ISOL (helium pressure/vapor isolation) switches (all four)—(closed) TANK ISOL switches all eight)—OP (open) CROSSFEED switches—CL.	
			07	Check aft LEFT (right) RCS (reaction control system): He PRESS switches—OP. TANK ISOL switches (all six) —GPC (hooked up to general-purpose computer) LEFT (RIGHT) RCS CROSSFEED switches—GPC MASTER RCS CROSSFEED—OFF.	
L–1:24:00	Orbit 17,300 mph (27,700 kph)	Auxiliary power unit (APU) prestart	R2	BOILER N_2 SPLY (boiler nitrogen supply) switches (all three)—ON. BLR CNTLR (boiler controller) switches (all three)—ON. BLR CNTLR PWR/HTR (power/heater) switches (all three)—position A. APU FUEL TK VLV (APU fuel tank valve) switches (all three)—CL.	Y: Control, this is *Columbia*, APU prestart complete, over. G: Roger, out.
L–1:21:00	Orbit 17,300 mph (27,700 kph)	Load deorbit computer program.	C2	On computer keyboard, enter: OPS 3 0 2 PRO.	
L–1:17:00	Orbit 17,300 mph (27,700 kph)	Go/no-go decision from Mission Control for deorbit, entry, and landing.			G: *Columbia*, this is Control. You are go for deorbit burn, over. Y: Roger, go for deorbit burn, out.

TIME	ALTITUDE/ VELOCITY	EVENT	PANEL	PROCEDURE	VOICE	5.7
L–1:15:00	Orbit 17,300 mph (27,700 kph)	Maneuver to deorbit burn attitude.	F6	Turn Orbiter around so it's flying tail first. FLT CNTLR POWER switch—ON.	Y: Control, this is *Columbia*. Maneuver to burn attitude complete, over.	
			F6, F8	On both panels, ADI ATT (attitude direction indicator [ADI]—attitude) switches—INRTL (inertial). ADI ERROR—MED. ADI RATE—MED. With rotation hand controller, maneuver to burn attitude. Compare attitude shown on CRT (cathode-ray-tube) with that on the ADI.	G: Roger, out.	
L–1:03:00	Orbit 17,300 mph (27,700 kph)	Single APU start.	R2	#1 APU FUEL TK VLV switch—OPEN. #1 APU CONTROL switch—START/RUN.	Y: Control, this is *Columbia*. We have single APU start, over.	
			F8	HYD PRESS (hydraulic pressure) indicator 1 should be LO green.	G: Roger, single APU start, out.	
			R2	HYD CIRC PUMP switches (all three)—OFF.		
L–1:02:00	Orbit 17,300 mph (27,700 kph)	Arm OMS engines.	C3	DAP (digital auto pilot) —AUTO MODE.	Y: Control, this is *Columbia*. OMS engines are armed, over.	
			08	L and R OMS He PRESS/ VAPOR ISOL A switches— GPC.	G: Roger, OMS armed, out.	
			C3	OMS ENG switches—ARM/ PRESS.		
L–1:00:15	Orbit 17,300 mph (27,700 kph)	Execute deorbit burn command on computer.	C2	On computer keyboard, push EXEC to begin countdown.		
*L–1:00:00	Orbit 17,300 mph (27,700 kph)	Deorbit burn.		The burn should last 2 to 3 minutes.	Y: Control, this is *Columbia*. Countdown to OMS burn ignition: five... four...three...two... one...ignition, over.	
					G: Roger, ignition, out.	
				Confirm when burn is complete.	Y: OMS burn complete, over.	
					G: Roger, burn complete, out.	

TIME	ALTITUDE/VELOCITY	EVENT	PANEL	PROCEDURE	VOICE
L–0:54:00	Orbit 17,100 mph (27,400 kph) (descending)	Post-OMS burn activities.	07 08	Check aft left and right RCS: He PRESS switches (all four)—OP. TANK ISOL switches (all six)—GPC. CROSSFEED switches (all four)—GPC. Check left and right OMS engine status: He PRESS/VAPOR ISOL switches (all four)—CL. TANK ISOL switches (all four)—OP. CROSSFEED switches (all four)—CL.	
L–0:52:00	Orbit 17,100 mph (27,400 kph) (descending)	Position Orbiter for entry.	02 C3	On computer keyboard, enter: OPS 3 0 3 PRO. Then enter: ITEM 2 4 (for roll) ITEM 2 5 (for pitch) ITEM 2 6 (for yaw). ORBITAL DAP—MAN (manual). Maneuver into position shown on CRT #1. Orbiter should face forward with the nose pointed up between 28° and 38°.	Y: Control, this is *Columbia*. We are in entry attitude, over. G: Roger, out.
L–0:50:00	Orbit 17,100 mph (27,400 kph) (descending)	Set and check switch positions for entry.	L2 L2/C3 C3 F6/F8 F8	CABIN RELIEF A and B—ENA. ANTISKID—ON. NOSE WHEEL STEERING—OFF. ENTRY ROLL MODE—OFF. Both Speed Brake/Throttle controls—FULL FORWARD. SRB SEP—AUTO. ET SEP—AUTO. AIR DATA—NAV. ADI ERROR—MED. ADI RATE—MED. HYD MAIN PUMP PRESS switches (all three)—NORM. HYD PRESS indicators (all three)—HI green.	Y: Control, this is *Columbia*. Entry switch checklist complete, over. G: Roger, out.

TIME	ALTITUDE/ VELOCITY	EVENT	PANEL	PROCEDURE	VOICE	5.9
L–0:41:00	Orbit 17,100 mph (27,400 kph) (descending)	Move Orbiter aero control surfaces to pre-pare hydraulic sys tem for entry and landing.	C2	On computer keyboard, enter: ITEM 3 9 EXEC. At L-36 minutes, stop the movement by entering: ITEM 4 0 EXEC.		
L–0:40:00	Orbit 17,100 mph (27,400 kph) (descending)	Dump propellants in the forward reaction control system over-board. This shifts the Orbiter's bal-ance point for entry.	C2	On computer keyboard, enter: ITEM 3 6 EXEC. (Item 36 arms the system.) Then dump the propellants by entering: ITEM 3 7 EXEC. Fifty seconds later, enter: ITEM 3 8 EXEC.	Y: Control, this is *Colum-bia*. RCS dump com-plete, over. G: Roger, out.	
L–0:35:00	Orbit 17,100 mph (27,400 kph) (descending)	Check entry attitude.	F6, F8	ADI should show: roll—0° pitch—28°–38° yaw—0°.		
		Inflate anti-G suit. Change computer pro-gram to next phase. Check switch positions.	C2 F2	Activation valve—ON. On computer keyboard, enter: OPS 3 0 4 PRO. Check following switches: SPDBK/THROT—AUTO PITCH—AUTO ROLL/YAW—AUTO BODY FLP—MAN.		
L–0:30:00	400,000 ft 17,100 mph (122,000 m/ 27,400 kph)	Atmospheric entry begins (entry inter-face).	F2	BODY FLP switch to AUTO.	Y: Control, this is *Colum-bia*. We are at entry interface, ready for LOS, over. G: Roger, out.	
*L–0:25:00	312,000 ft 16,700 mph (95,100 m/ 26,800 kph)	Begin communications blackout caused by ionized particles enveloping Orbiter at entry.				
		RCS roll thrusters deactivate.		When flight instruments detect pressure of 10 psf (pounds per square foot) (479 pascals), roll thrusters are turned off. Elevons now control roll.		

TIME	ALTITUDE/ VELOCITY	EVENT	PANEL	PROCEDURE	VOICE
L–0:23:00	15,000 mph (24,000 kph)	RCS pitch thrusters deactivate.		When flight instruments detect pressure of 20 psi (958 pascals), pitch thrusters are turned off. Elevons now control pitch.	
*L–0:20:00	230,000 ft 15,000 mph (70,100 m/ 24,000 kph)	Maximum heating— Nose and wing leading edges reach 1,500°C (2,800°F).			
L–0:16:00		First roll reversal (S-turn).	F6 F2	ADI RATE switch—HI. ROLL/YAW switch—CSS (control stick steering). Use rotational hand controller (RHC) for the maneuver.	
L–0:12:00	180,000 ft 8,300 mph (55,000 m/ 13,000 kph)	Leave communications blackout.			Y: Control, this is *Columbia*. Do you copy? Over. G: We copy, *Columbia*. You look good, over. Y: All systems go, out.
		Second roll reversal (S-turn).		Move RHC in opposite direction from last S-turn.	
L–0:10:00		Speed brake to 100%.	L2 F7	Move speed brake handle back to 100%. Check SPEED BRAKE indicator; it should show 100%.	
L–0:07:00		Third roll reversal (S-turn).		Move RHC opposite direction.	
L–0:06:00	90,000 ft (27,000 m) Mach 3.3 (3.3 times the speed of sound)	Move speed brake to 65% Deploy air data probes.	L2 C3	Move speed brake handle to 65%. AIR DATA PROBE switches (both)—DEPLOY	

| --- | --- | --- | --- | --- | --- | --- |
L—0:05:30	83,000 ft (25,000 m) Mach 2.5	Fourth roll reversal (S-turn).		Move RCH opposite direction.		
		Begin terminal area energy management (TAEM). Head for heading alignment cylinder Waypoint One to line up on runway for final approach.			Y: Control, this is *Columbia*. TAEM Interface. Everything looks fine, over. G: Roger, out.	
*L—0:03:00	50,000 ft (15,000 m) Mach 1.0	RCS yaw thrusters deactivate.		Use rudder via pedals for yaw control.		
L—0:02:00	13,300 ft 424 mph (4,050 m/ 678 kph)	Begin autoland guidance.	F2	CSS switches for PITCH and ROLL/YAW—ON. You now have full manual control. Use RHC and speed brake as required to hold a 22° glideslope.	G: *Columbia*, this is Control. Have you acquired autoland? Over? Y: Roger, out.	
L—0:00:30	2,000 ft 350 mph (600 m/ 560 kph)	Begin preflare.		Using RHC and speed brake, adjust glideslope to 1.5°.	Y: Control, this is *Columbia*. Preflare initiated, over. G: Roger, out.	
L—0:00:17	135 ft 340 mph (41 m/ 540 kph)	Complete preflare. Arm landing gear.	F6	LANDING GEAR ARM switch—lift cover and push.		
L—0:00:14	90 ft 330 mph (27 m/ 530 kph)	Deploy landing gear.	F6	LANDING GEAR DN (down) switch—lift cover and push.	G: *Columbia*, this is Control. Gear down, over. Y: Roger, gear down and locked. Out.	
L—0:00:00	0 ft 215 mph (340 kph)	Touchdown.	L2 C3	Speed Brake to 100%. Using RHC, pitch forward to lower nose, then full forward when nose wheel touches down. Using pedals, brake as required. SRB SEP switch—MAN/AUTO. ET SEP switch—AUTO. Lift covers on ET and SRB SEP buttons and PUSH.	G: *Columbia*, main gear at ten feet . . . five feet . . . four feet . . . three feet . . . two feet . . . one, contact. Nosewheel at five feet, four , three, two, one, contact. Y: Roger, out.	

TIME	ALTITUDE/VELOCITY	EVENT	PANEL	PROCEDURE	VOICE
L + 0:02:00	0 ft 0 mph	Orbiter stop.	L2	Speed brake full forward.	Y: *Columbia* to Convoy 1, "Wheels Stop," over. G: This is Convoy 1, "Wheels Stop," out.
		Turn off APUs.	R2	APU AUTO SHUT DN switches—ENA.	
L + 0:04:00		Deactivate OMS and RCS systems to prepare to turn Orbiter over to the ground crew.	C3	OMS ENG switches (both)—OFF.	
			07	Set the following switches: For L and R aft RCS: MANIFOLD ISOL (All ten)—OP TANK ISOL (all six)—OP He PRESS (all four)—CL CROSSFEED (all four)—CL.	
			08	For L and R OMS: He PRESS/VAP ISOL (all four)—OP CROSSFEED (all four)—CL. For forward RCS: MANIFOLD ISOL (all five)—OP TANK ISOL (both)—OP He PRESS (both)—OP	
			C2	On computer keyboard enter: OPS 9 0 1 PRO.	
L + 0:27:00		Leave the Orbiter— YOU ARE HOME.		Leave the Orbiter. Enter the egress vehicle.	Y: Control, this is *Columbia*. We are "ready for egress," over. G: Convoy 1, this is Control. "Proceed with crew egress," over. G: Control, this is Convoy 1. Roger, out.

**The Space Shuttle will
fly many different missions.
Here are five that you can fly.**

Crew: **Commander, pilot, mission specialist, payload specialist**

Duration: **3 days**

Launch Site: **Kennedy Space Center**

Payload: **Multimission Modular Spacecraft support system, remote manipulator arm, manned maneuvering unit.**

Altitude: **300 miles (480 kilometers)**

On February 14, 1980, NASA launched the Solar Maximum Mission (SMM) into space with an expendable Delta launch vehicle. This satellite was the first to use the Multimission Modular Spacecraft and was equipped with a manipulator-arm grapple so it could be retrieved by the Shuttle. (Review 3.28· for a description of the Multimission Modular Spacecraft.)

SMM was designed to provide high-resolution, precisely aimed images of the sun during the peak of its 11-year sunspot cycle. In the first months after launch, scientists observed the sun with SMM's instruments. In late 1980, however, the satellite's control system failed when three fuses blew. Since that time, the spacecraft has been slowly tumbling out of control in space. In fact, it is tumbling too fast for the Shuttle's manipulator arm to latch onto it.

However, you can salvage the $75-million satellite and return it to useful service. SMM is tumbling too fast for the manipulator arm, but not so fast that an astronaut wearing a manned maneuvering unit couldn't bring it under control. Once you've stopped the satellite from tumbling, your crewmates inside the Shuttle can grab it with the manipulator arm and bring it into the cargo bay. There, you can repair the ailing satellite. To enable the Orbiter to reach the SMM, the only payloads you'll carry are the manipulator arm, a platform to hold the satellite in the cargo bay, a replacement attitude control module, and two manned maneuvering units.

Turn to 1.18 for launch and ascent. Once you reach an orbit of 300 miles (480 kilometers) and the cargo bay doors are open, proceed with the mission. Computer program codes 201 and 202 will give you specific instructions for maneuvering to the satellite. These maneuvers will take most of your first day in space. After reaching the SMM, fly around the satellite and inspect it. You should find it rotating about its long axis at a rate of about 0.9° per second, or nearly one revolution every 6½ minutes. When you've completed the inspection, maneuver the Orbiter to a position about 1,000 feet (300 meters) from SMM. Report what you saw on your fly-around inspection to Mission Control, then go ahead and prepare dinner and relax. Try to get a good night's sleep—tomorrow will be a very busy day.

On mission day two, you'll get up an hour earlier than usual to have more time to prepare for the day's activities. When you've finished breakfast and cleaned up, begin your pre-EVA activities. Two hours before you begin your spacewalk, put on the liquid-cooling and ventilation garment, and begin breathing pure oxygen with the portable oxygen system (POS). After breathing oxygen for two hours, you'll have flushed all the nitrogen from your blood stream and you can put on your spacesuit. One of your fellow crewmembers will go along on the EVA in case you need help.

Transfer to the mid-deck and enter the airlock to begin donning your spacesuit. Spacesuit-donning procedures are shown in 3.10 When the airlock is depressurized, open the outer hatch and, using a hand-over-hand movement along the handrails, move over to the manned maneuvering unit. The unit is in a cradle in the front of the cargo bay. You'll also have a special plate that attaches to the chest of your spacesuit. This plate allows you to dock with SMM's grapple. When you're in the maneuvering unit and have the plate on your chest, you can maneuver over to the satellite. Match the spacecraft's rotation, then close with the grapple. Then use the maneuvering unit to control the SMM. Undock and move around to the opposite side of the satellite so you can steady it while the manipulator arm moves it into the cargo bay.

There, you'll remove two large screws on the attitude control module. Removing these screws disconnects all mechanical and electrical connections between the module and the rest of the satellite. A special tool has been devised for you to use when removing the screws. If you used a regular screwdriver in weightlessness, you'd probably turn while the screw remained unmoved.

After the faulty attitude control module has been removed, replace it with the good one in the cargo bay. Work slowly and carefully while moving these around. They're big—47 by 47 by 18 inches (1.2 by 1.2 by 0.5 meters)—and weigh several hundred pounds on earth. Even though they are weightless in orbit, they still have **inertia** when set in motion. It's not easy to stop something that big once you start it moving.

Your crewmates inside the Shuttle will send a series of commands to SMM to check its operation after you've finished. The next step is to return the satellite to orbit and make one last check to make sure it's working.

Return the maneuvering unit to its cradle and reenter and repressurize the airlock. After you've stowed your space suit, go into the mid-deck and have a much-deserved (and delayed) lunch.

Early on day three, return to the Kennedy Space Center, using the checklist given in 5.5

Satellite Repair Checklist

TIME (hours: minutes)	ACTIVITY
EVA—2:20	Transfer to mid-deck, put on liquid-cooling and ventilation garment and urine collector.
EVA – 2:10	Transfer to Flight Deck and unstow Portable Oxygen System (POS). Verify that POS pressure is 3,270 psi (22,500 kilopascals).
EVA – 2:05	1. Connect POS oxygen hose to POS O_2 QD (POS oxygen quick-disconnect fitting) on panel C6.
	2. On panel C6, POS 2—OPEN.
	3. Depress POS PURGE BUTTON for 5 seconds to purge mask and re-breather loop with oxygen.
EVA – 2:00	1. Don POS mask and start prebreathing oxygen.
	2. Connect POS straps, wearing unit around your waist.
	3. Continue activities in flight deck.
EVA – 1:00	1. On POS pack, POS O_2 switch to ON position. This changes your oxygen source from the Orbiter to the POS pack.
	2. Disconnect POS oxygen hose from panel C6, wrap it around your waist.
	3. On Panel C6—POS 2 switch to CLOSE.
	4. Transfer to middeck.
	5. Unpack a spare POS to take into airlock with you.
	6. Open inner airlock hatch.
	7. Turn on airlock floodlights. (switches on panel MO13Q).
	8. Enter airlock
	9. Unwrap oxygen hose from around your waist and connect it to EV 2 POS O_2 QD (EVA 2 POS oxygen quick-disconnect fitting) 109. On Panel AW82A in airlock, EV 2 EMU O_2 SPLY to OPEN (EVA 2 spacesuit oxygen supply to open). This again supplies oxygen from the Orbiter to your POS 11. On POS pack, POS O_2 to OFF.

EVA – 0:45	1. Begin inspection of spacesuit. Make sure you have your helmet, gloves, in-suit drinking bag, and EVA procedures checklist.
	2. Install the drinking bag in the suit's upper torso. This gives you something to drink while you're working in space.
	3. Clip the EVA procedures checklist to the suit's left wrist.
EVA – 0:30	1. Put on lower torso.
	2. Unpack spare POS pack. You'll use this one with a mouthpiece while donning the suit.
	3. Turn on spare POS.
	4. Take a deep breath and **hold**.
	5. Take off mask, install mouthpiece, put on nose clip, and begin breathing through your mouth.
	6. On panel AW82A, EV 2 EMU O_2 SPLY switch to CLOSE.
	7. Stow the POS you were using.
	8. Put thumb loops on your cooling garment over your thumbs.
	9. Take a deep breath and **hold**.
	10. Take mouthpiece out of your mouth, take off nose clip.
	11. While still holding your breath, wriggle up into the upper torso.
	12. Install mouthpiece and nose clip, resume breathing.
	13. Attach liquid cooling garment to umbilical from life-support backpack.
	14. Connect upper and lower torsos. As you close the waist ring, you should feel and hear a series of faint clicks.
	15. Position the in-suit drinking bag's valve so you can reach it with your mouth.
	16. Put on communications carrier ("Snoopy hat") and connect it to suit communications umbilical.
	17. Release right thumb loop, put on right glove, lock glove.
	18. Release left thumb loop, put on left glove, lock glove.
	19. On your check pack, set sliding oxygen control switch to PRESS (pressurize) position.
	20. Place helmet over your head.
	21. Hold your breath, remove the POS

mouthpiece and nose clip, lock the helmet on the suit neck ring. Again, you should hear a series of faint clicks as the helmet locks into position.

22. Watch your suit-pressure gauge. Pressure in the suit should build up to 4 psi (28 kilopascals) above ambient conditions.
23. Move sliding switch on your chest pack from PRESS to OFF. Chest pack display will indicate if there is any leakage.
24. Move sliding switch on chest pack to EVA position.
25. Disconnect suit from wall mounted frame.
26. Close inner airlock hatch.
27. On airlock panel AW 82A, turn AIRLOCK DEPRESS switch to 5.
28. Turn AIRLOCK DEPRESS switch to 0, then wait 3 minutes.

EVA − 0:0
1. Open outer airlock hatch and emerge into the cargo bay.
2. Connect tether to your suit and airlock anchor.

EVA + 0:05
1. Move over to the manned maneuvering unit (MMU).
2. Visually inspect MMU.
3. Attach satellite grapple plate to your chest pack.

EVA + 0:15
1. Back into MMU and latch it to suit backpack.
2. Adjust MMU control arms from launch to flight position.
3. Fire the -X thrusters to check the MMU propulsion system.
4. Release the MMU from its cradle.
5. Push upwards with your feet to start movement away from the cradle.
6. Translate upward, stop, perform a 180° turn (rotation) to the right, stop, rotate 180° to the left, stop, pitch up 90°, stop, pitch down 90°.

EVA + 0:30
1. Translate over to the Solar Maximum Mission Satellite (SMM).
2. Release your tether to the airlock.
3. Match SMM rotation, translate closer in to the satellite, capture grapple with your chest plate.
4. Using maneuvering unit thrusters, stop SMM's rotation. Bring it into position so the Orbiter's manipulator arm can reach grapple.

EVA + 1:00
1. Release your chest plate from SMM grapple.
2. Move over to other side of satellite.
3. Steady SMM while the payload arm captures the grapple.
4. Steady satellite while manipulator arm moves it into the cargo bay and onto the SMM work platform.
5. Release satellite once it's in the cargo bay.

EVA + 2:00
1. Attach a tether between yourself and the SMM work platform.

EVA + 2:10
1. Begin to repair satellite.
2. Identify attitude control module on satellite.
3. Unpack large screw-removing tool from tool box on work platform. Tether it to your wrist.
4. Remove two large screws at top and bottom of attitude control module.
5. Move module into the storage space on the platform and secure it.

EVA + 3:30
1. Unpack replacement module and move it into position on SMM.
2. Attach large screws at top and bottom of attitude control module.

EVA + 4:30
1. Re-stow screw removal tool in tool box on platform.
2. Release your tether to platform.
3. Move back away from SMM, reattach tether to airlock, which should be floating nearby.
4. Stand by to assist if necessary while manipulator arm moves SMM out of cargo bay.

EVA + 5:00
1. Once SMM has been released, fly over to the MMU cradle and back into it until unit latches into its storage position.
2. Turn unit off, fold up armrests.
3. Release your suit from the maneuvering unit.
4. Remove and stow chestplate.

EVA + 5:15
1. Move from MMU back into airlock.
2. Secure airlock outer hatch.
3. Repressurize airlock.
4. Doff spacesuit, replace it on its storage rack in airlock.
5. Open airlock inner hatch and rejoin rest of crew in Orbiter.

Crew: **Commander, pilot, mission specialist**
Duration: **3 days**
Launch Site: **Kennedy Space Center**
Payload: **Two communications satellites equipped with Payload Assist Modules**
Altitude: **200 miles (320 kilometers)**

On this mission, you'll transport two communications satellites into space. After you've released them, strap-on rockets propel them to geostationary altitude. One of the satellites was built by a foreign government and will be their first step toward establishing a national communications system. The other, built by an American firm, will relay business data across the United States. Each satellite is attached to a payload assist module-D (PAM-D). Before the Space Shuttle was built, satellites like these were launched on separate rockets, at a cost of $25 million a piece. A single Shuttle flight costs $35 million. This means a launch cost of $17.5 million per satellite on this mission—a savings of 30%.

For launch and ascent, go back to 1.18. When you're in a circular orbit 200 miles (320 kilometers) high and the cargo-bay doors are open, proceed with the mission.

Release the satellites one at a time on the first and second days of your flight. For launch, each satellite with its attached upper-stage motor is supported in the cargo bay by an aluminum cradle. The satellite PAM sits on a tilting "spin table" in the cradle.

This table has two functions—first it raises the satellite into release position, then it spins the payload for stability. During ascent, the payload's long axis is parallel to the Orbiter's. For release, it elevates so its long axis is at 45° to the Shuttle's. Once the satellite is in position, two electric motors spin the table like a record turntable. Propellants in small solid motors like the PAM sometimes burn unevenly, resulting in an off-center thrust that can push the rocket off course. Spinning such small solid-propellant motors cancels the effects of any thrust imbalances, ensuring an on-course flight. Once you've verified that the spin table is up to speed, prepare to release the satellites. During these last minutes before release, check the satellite one last time to make sure it is operating properly. When you release the satellite,

springs push it away from the Orbiter. As the two craft (satellite and the Orbiter) drift apart, their maximum separation occurs one-half orbit after release, or at T + 45 minutes. At that point, a timer on the PAM ignites the motor. One minute, twenty-eight seconds later, the motor's propellants are consumed. Explosive bolts fire, and the satellite separates from the spent PAM. The satellite coasts up to an altitude of 22,300 miles (35,800 kilometers). When it reaches this altitude, a small built-in rocket, called an "apogee kick motor," fires to place the satellite in its geostationary orbit.

The next day, repeat the procedure for the other satellite.

Early on mission day three, return for landing at the Kennedy Space Center using the checklist in 5.5

PAM-D Release Checklist

TIME (minutes:seconds)	EVENT
T—25:00	Turn on PAM power; check system status.
T—20:00	Status check complete.
T—15:00	Maneuver Orbiter into position for release.
T—14:00	Unlock spin table; satellite lifts into deployment position.
T—8:00	Start spin table.
T—6:50	Spin-up complete.
T—0:00	Release satellite-PAM combination.
T + 45:00	PAM ignition.
T + 46:28	PAM burnout.
T + 48:00	Fire PAM-satellite separation bolt cutters; satellite coasts to geostationary altitude.

Crew:　　　　**Commander, pilot, mission specialist**
Duration:　　**4 days**
Launch Site:　**Kennedy Space Center**
Payload:　　　**Space Telescope, remote manipulator arm**
Altitude:　　　**300 miles (480 kilometers)**

This is one of the most important missions in the entire shuttle program. You will not only deploy the Space Telescope, the most powerful optical tool astronomers have ever had, but you will also retrieve the Long-Duration Exposure Facility (LDEF). The LDEF experiments will play a key role in design of elements for long term space flight. (See 3.25 and 3.26 for additional payload information.)

Use 1.18 for launch procedures and ascent. After you have achieved orbit at 300 miles (480 kilometers) and the cargo-bay doors are open, you are ready to proceed.

For the rest of day one, you will be performing checkouts of the space telescope systems to be sure they are in satisfactory shape after the launch. You will also run through tests of the remote manipulator system (RMS) to be sure that it is functioning. You should also review Free-Flying Payload Deployment in 3.8

On day two, after routine breakfast and housekeeping duties, the pilot and mission specialist will go to the aft crew station. The RMS is powered up, set to IDLE mode, and the shoulder brace released. Deploy RMS, release the retention latch, and select the joint angle. Turn on the cargo-bay lights, and select the closed-circuit TV monitors.

Activate the Telescope systems and repeat the system checks of Telescope power circuits, motor circuits, detection instrument, gyros, and telemetry systems.

Set coordinates for X-, Y-, and Z- axis translation and for pitch, yaw, and roll for the arm end effector. Maneuver the Orbiter to the correct attitude using the digital autopilot (DAP, see 1.29). Once this is done, move the arm to the grapple fixture and grapple the Telescope. Release the holddown latches on the Telescope and remove it from its mountings. Now enter the coordinates for deployment into the computer. Once the Telescope is properly oriented, and any vibration is damped out, release the end effector and carefully return the arm to its cradle.

Carefully maneuver the Orbiter away from the Telescope and observe the deployment of the solar panels and the ground-controlled checkout. Get to bed early, because day three will be busy.

The third day will have two major activities: orbital transfer and LDEF retrieval. The LDEF is orbiting about 50 miles (80 kilometers) above you, so the orbital maneuvering system (OMS) must be fired to increase the altitude of your orbit. Power up the OMS, and use the DAP to maneuver the Orbiter to the correct burn attitude. The first burn will put you in an elliptical orbit with a 350-mile (560-kilometer) apogee. Another short burn will circularize this orbit and you will be flying parallel to the LDEF. The reaction control system will now maneuver you to a rendezvous, and orient the Orbiter for retrieval.

Power up and unlatch the RMS again, and move the end effector to grapple the fixture on the LDEF. Move the LDEF carefully into the cargo bay. Once you have cradled and latched the LDEF, remove the end effector and return the arm to its stowed position.

On day four, you will deorbit, enter, and land the Orbiter using the checklist in 5.5

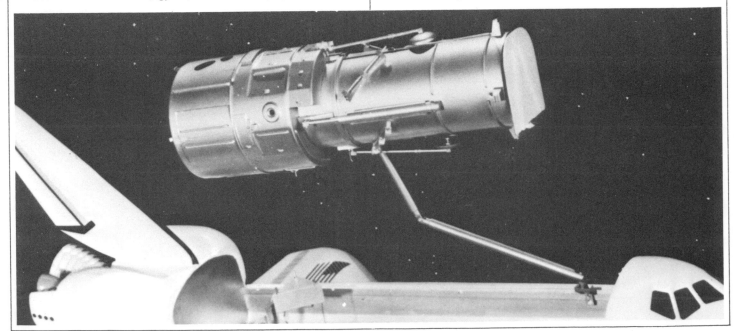

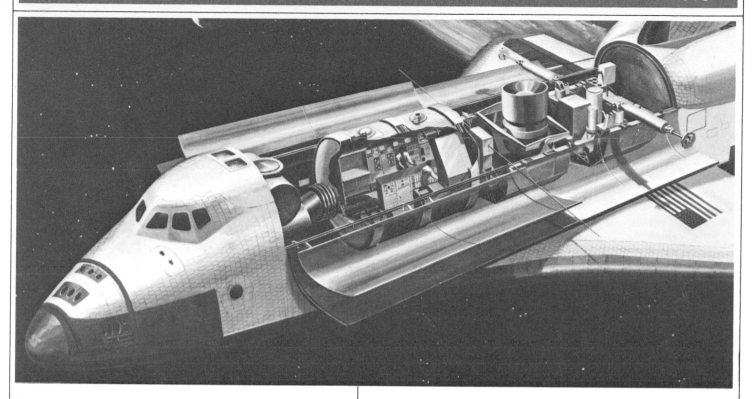

Crew: **Commander, pilot, mission specialist, two payload specialists**

Duration: **5 days**

Launch Site: **Kennedy Space Center**

Payload: **Spacelab crew module, one pallet, two small self-contained payloads**

Altitude: **150 miles (240 kilometers)**

Spacelab, as you will recall from 3.21, is a flying laboratory composed of pressurized cylindrical crew modules and unpressurized instrument-carrying pallets. The laboratory remains in the Orbiter cargo bay throughout the flight. Payload specialists transfer to the crew module via a pressurized tunnel attached to the Orbiter mid-deck.

On this mission, you'll investigate the effects of microgravity on materials and living organisms. Additionally, you have two small self-contained payloads on board.

During launch and ascent, the two payload specialists sit in the Orbiter mid-deck. The other three crew members' seats are in the flight deck. Use 1.18 for launch and ascent. When you're in a circular orbit 150 miles (240 kilometers) high and the cargo-bay doors are open, proceed with the mission.

One of your first activities will be to open the hatch leading to the Spacelab transfer tunnel, and to move into the crew module. Check all the electrical equipment and the laboratory animals caged in the module. You will find mice, gerbils, fish, and monkeys in your "flying zoo." The animals may be somewhat agitated and restless, but they'll soon calm down. After you've

completed the module checkout, conduct a test of the pallet experiments. Once you're finished, crawl back through the tunnel into the mid-deck and have lunch.

During your first afternoon in space, you'll begin an experiment to observe mobility and adaptation to microgravity. This experiment is a space-age variant of the mouse-in-a-maze experiment, only the maze you'll use is three-dimensional. The maze is made of plastic tubes. Before launch, the mice you have aboard were placed in a similar device and the times it took them to complete the maze were recorded. Follow the same procedure; that is, place them, one at a time, in the entrance and record how long it takes each mouse to complete the course. Repeat the experiment using the gerbils when you're finished with the mice. This should take most of the afternoon. While you're doing this, one of the payload specialists will calibrate the cosmic-ray counters mounted on the pallet.

On your second day in space, you'll perform several materials-processing experiments and repeat the maze tests with the mice and gerbils. Also, just after breakfast, turn on the small self-contained payloads. One of them consists of a device to photograph the growth of ice crystals in microgravity. This data on how microgravity affects the formation of crystals may lead to important breakthroughs in manufacturing techniques in space. The other small payload contains a device to deposit a thin coat of metal on thread. Such coated threads can be used to make extremely strong fiber-wound pressure vessels. In this particular experiment, nickel is vaporized and deposited on glass thread. After the flight is over, the coated thread will be examined to see if the process works better in space than it does on earth.

Your first materials-processing experiment in the crew module involves the growth and formation of large crystals of germanium and silicon in zero-g. You'll place samples of various compounds in an oven, melt them, and allow them to cool. These samples will be compared to similar ones processed on the ground to observe the effects of microgravity on them. Experiments such as these are very important to the electronics industry—they will determine if large crystals for electrical devices can be grown in space.

While you're doing these tests, the two payload specialists will be busy. The first, a physicist (who calibrated the cosmic-ray counters yesterday), will be working at his console, measuring cosmic rays, and pointing other sensors on the pallet at the sun. His observations, which will be made every day of the flight, may help unlock the secrets of the sun and enable us to understand how to tame fusion reactions for power generators on earth. The second payload specialist is a veterinarian. Her tasks are to draw samples of blood and bone marrow from the monkeys and to observe how the absence of gravity affects the way the fish swim. This second experiment is relatively simple—all she has to do is set up a camera in front of the fishtank and film the fish as they swim. It will provide useful data on the function of our vestibular, or balance-sensing, organs.

Your materials-processing tests will take most of the morning. When you've finished, prepare lunch for the crew. After lunch, go back to the module and repeat the maze experiment from the previous day. Repeating the experiment each day should yield data on how small mammals adapt to weightlessness.

The veterinarian payload specialist has another experiment to perform this afternoon, and you and the other payload specialist will be test subjects. This is a repeat of a Skylab experiment first performed in 1973, the rotating litter chair. You'll sit, blindfolded, in a chair that spins. She'll tell you to put your head in different positions as you spin to see if you get dizzy. Like the fish experiment, this measures the effect of zero-g on your vestibular organs.

Your third day involves a test of a process called **electrophoresis**. All living cells have a small negative electrical charge on their surfaces. Different cells differ in the amount of the charge, their size, and their shape. You can separate cells by using their different charges. On earth, this is difficult because the electrical charges are extremely small and gravity causes sedimentation and convection. In microgravity, however, cells in solution can be separated by electrophoresis. You'll be separating different kinds of kidney cells to extract the enzyme urokinase. This enzyme is used to dissolve blood clots. It can be extracted on earth, but it is very difficult to do so. Electrophoresis in space should enable you to process this material easily and cheaply. As you've done on your last two days, repeat the maze experiment in the afternoon. Also, repeat the rotating chair experiment.

Your last full day in space will be occupied with experiments in the morning and clean-up in the afternoon. After breakfast, turn off the two small self-contained payloads and transfer to the Spacelab crew module. Perform the maze and rotating chair experiments one last time, then break for lunch. Spend your afternoon putting away all the equipment in the crew module. Make sure the animals have plenty of food and water, crawl back through the transfer tunnel, and close the hatch to Spacelab.

On day five, have breakfast, unstow the seats for the mission and payload specialists, and return to earth using the checklist in 5.5

Earth Resources Satellite Deployment Mission

Crew: **Commander, pilot, mission specialist**
Duration: **7 days**
Launch Site: **Vandenberg Air Force Base**
Payload: **Earth Resources Satellite, Spacelab pallet-mounted earth magnetosphere observation package**
Altitude: **575 miles (920 kilometers)**

From the vantage point of space, a bird's-eye view of earth is available. A satellite in orbit over the poles can monitor the entire planet every 18 days. Equipped with capable sensors, the satellite surveys land area, ocean area, and atmospheric changes, and can monitor the health and location of forests and crops. Various types of pollution can be detected, and planning of land use and water access, and even exploration for petroleum and minerals is being done.

With new sensors being developed and many of the nations on earth growing to rely on this information, your deployment of this Earth Resources Satellite is very important. To best do its job, the satellite will be placed in polar orbit. Launches from the Kennedy Space Center in Florida go eastward and are equatorial. Launches from Vandenberg Air Force Base at the Western Test Range near Lompoc, California, go southward—over the poles.

The third section of the cargo bay contains a Spacelab pallet. Your polar orbit is ideal for studying the earth's magnetic field, its shape, structure, and how it interacts with the atmosphere. The earth-magnetosphere instrument package on the pallet consists of instruments to study this. A magnetometer, charged-particle detectors, ultraviolet detectors, and a Geiger counter will all be used. You will, however, return from the satellite-deployment altitude (575 miles, 920 kilometers) to a lower orbit at 200 miles (320 kilometers) to avoid the radiation hazard of prolonged stay in the Van Allen radiation belt.

The ground facilities are somewhat different in California (review 7.31, Vandenberg), but the Orbiter systems and basic launch sequence are identical. Use 1.18 for launch and ascent.

Once the cargo doors are open, you are ready for orbital operations. On day two, you are going to perform a critical orbital transfer maneuver. You will go from your parking orbit at 175 miles (280 kilometers) high to one 575 miles (920 kilometers) high. To accomplish this, you will need an Orbital Maneuvering System (OMS) kit. The OMS can operate from auxiliary fuel tanks located in the cargo bay. For this mission, the OMS kit will take up about a third of the cargo bay, with the spacecraft and experiment pallet in the remaining 40 feet (12 meters).

The additional fuel is sufficient to carry you to the higher orbit, perform the necessary maneuvers, and bring you down for entry and landing. The computers will have already been programmed for the two OMS burns you will require.

Once you are parked in orbit at 575 miles (920 kilometers), you are in the lower Van Allen belt. You cannot spend long periods at this altitude because of the radiation hazard. On day three, you will prepare to deploy the satellite. First, check the systems on the spacecraft —test the batteries, power system, and the sensory, guidance, and telemetry systems. Next power up the remote manipulator arm. Light the cargo bay and select the TV cameras and viewing monitors (see 3.4).

Using the arm, grapple the spacecraft. Unlatch the satellite. Then enter coordinates for translation and rotation. The computer will then perform the removal and deployment. Once you have damped any wobble in the satellite, release the end effector and stow the arm. Move the Orbiter away.

Ground controllers will now perform an evaluation of the spacecraft.

Upon completion of the deployment, you will reposition the Orbiter to transfer to a lower orbit. This maneuver is similar to the one you will make for atmosphere entry, so you can use the digital autopilot/ RCS information and OMS firing procedures from 5.5. A small correction burn will be necessary when you have reached your lower orbit. It's been a long day, so an extra-long rest period is scheduled for day four.

On the fourth day, after your breakfast, housekeeping, and check of the Orbiter systems, you will power up the instrumented pallet using the power switch panels at the aft crew station. You should visually inspect the pallet through the window for any obvious damage from yesterday's operations.

Electrical tests are then performed on the power supplies, the instruments, flight-data recorders, and the telemetry downlinks. These carry information to scientists on the ground. During the experiments, they may recommend changes to increase the value of the data you are getting.

You will monitor the pallet, which will operate continuously for three days.

Also stowed aboard are about half a dozen experiments supplied by students and scientists from around the world. You will be asked to perform them, videotape the results, and stow them for return to earth.

Once you have completed your experiments on day seven, use the entry and landing procedures in 5.5.

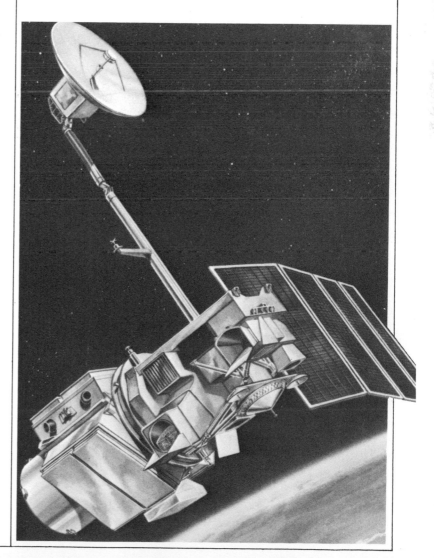

6.12

At launch, the Space Shuttle comprises three major subassemblies: the Orbiter, the External Tank, and two Solid Rocket Boosters. The vehicle is 184 feet tall overall, weighs 4.4 million pounds, and has a thrust of over 6 million pounds.

Technical details for the Orbiter, tank, and boosters are given in the next several sections. A list of the principal contractors used in the construction of the Shuttle gives you an indication both of the remarkable technology involved and of the widespread participation by men and women in firms all over the U.S.A. whose expert contribution is built into this extraordinary air-space vehicle.

The Space Shuttle Orbiter is one of the most remarkable flying machines ever built. About the size of a DC-9 airliner, it is the crew- and payload-carrying part of the Shuttle. Most of the Orbiter's fuselage is occupied by the 15-by-60-foot (4.5-by-18-meter) cargo bay, which can carry payloads weighing up to 65,000 (29,500 kilograms) pounds. Besides this enormous cargo, the Orbiter can accomodate up to seven people on flights lasting up to 30 days.

Overall Dimensions

Length:	122 feet (37 meters)
Height:	57 feet (17 meters)
Wingspan:	78 feet (24 meters)
Weight (empty):	165,000 pounds
	(75,000) kilograms)

Structure

The structure is divided into five sections: the forward fuselage, mid fuselage, aft fuselage, vertical stabilizer, and wings. Most of the structure is made from aluminum. Titanium and a boron-epoxy composite material are used in the aft section. Such lightweight high-strength materials are needed for the main-engine thrust structure.

The **wings** and **vertical stabilizer** have an airplane-style aluminum framework of ribs and spars. Elevons on each wing control pitch and roll during atmospheric flight. Yaw is controlled by a movable rudder on the stabilizer. The rudder also splits in half to act as a speed brake.

The **mid fuselage**, the center portion of the Orbiter, contains the cargo bay. Many other important subsystems are housed in the mid fuselage. Three fuel cells, which generate electricity, are housed beneath the cargo-bay floor. In addition, hydrogen and oxygen for the fuel cells and oxygen and nitrogen for the cabin atmosphere are stored there.

Two large doors cover the cargo bay. These doors are made from a graphite-epoxy composite material. They are the largest such composite structures built for flight use.

Your crew compartment is in the **forward fuse-lage**. Machined aluminum panels are welded together to make the pressurized compartment. Aluminum frames attached to the compartment give the Orbiter its outward shape. This area also houses the forward landing gear and the pod for the forward reaction control system.

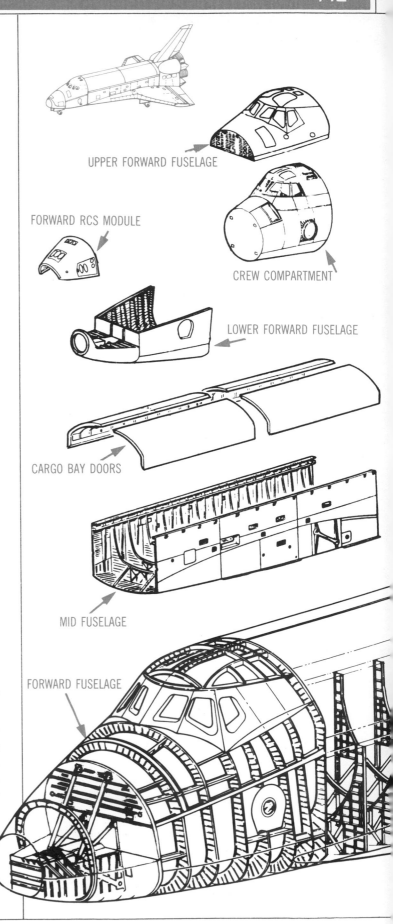

UPPER FORWARD FUSELAGE

FORWARD RCS MODULE

CREW COMPARTMENT

LOWER FORWARD FUSELAGE

CARGO BAY DOORS

MID FUSELAGE

FORWARD FUSELAGE

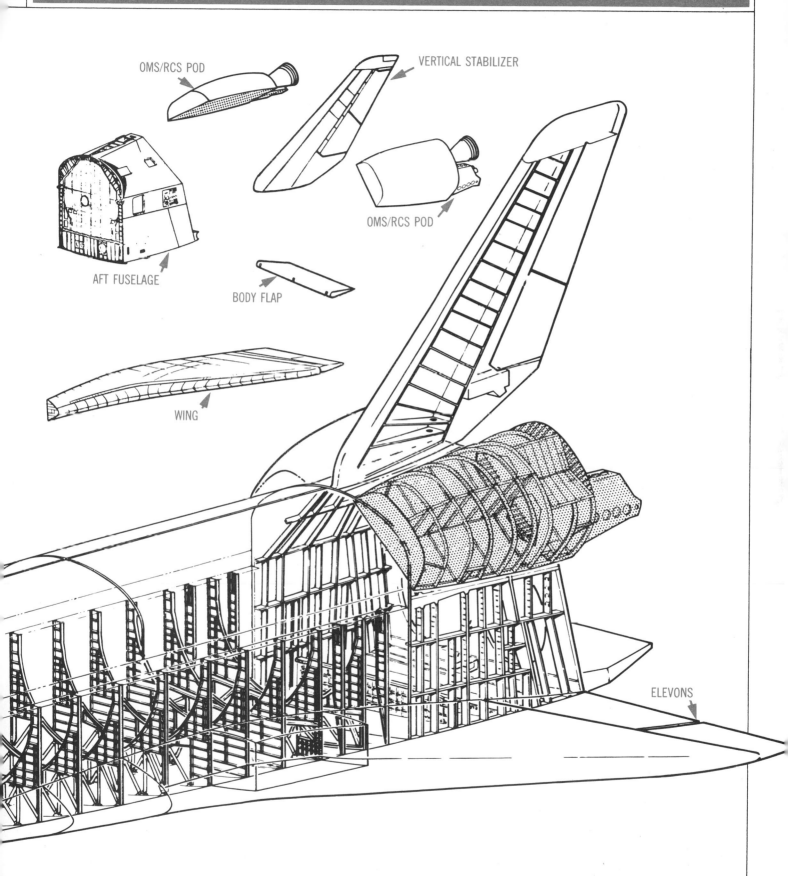

OMS/RCS POD

VERTICAL STABILIZER

AFT FUSELAGE

OMS/RCS POD

BODY FLAP

WING

ELEVONS

STRUCTURE—LEFT SIDE

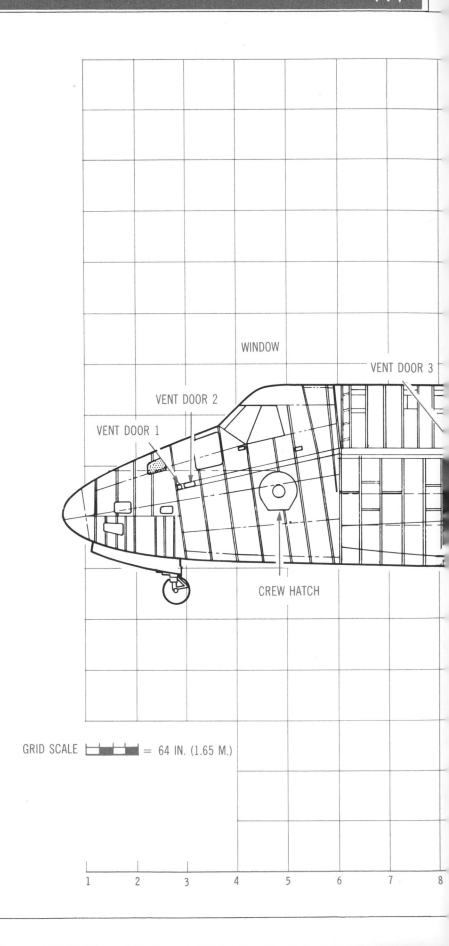

WINDOW

VENT DOOR 3

VENT DOOR 2

VENT DOOR 1

CREW HATCH

GRID SCALE ⊢─■─■─┤ = 64 IN. (1.65 M.)

1 2 3 4 5 6 7 8

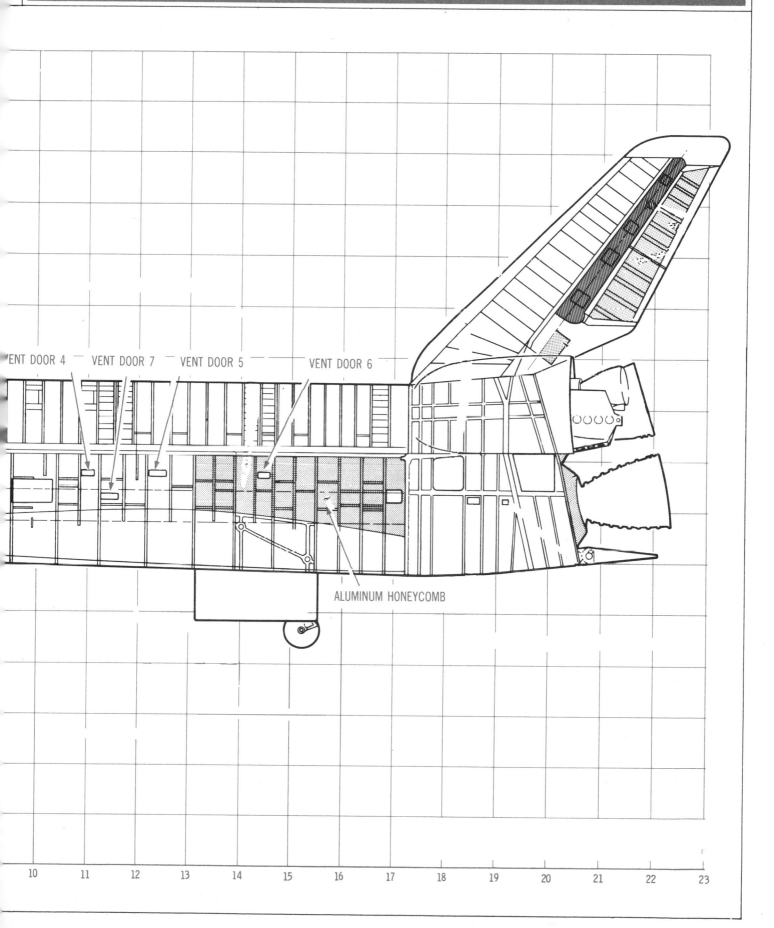

VENT DOOR 4   VENT DOOR 7   VENT DOOR 5   VENT DOOR 6

ALUMINUM HONEYCOMB

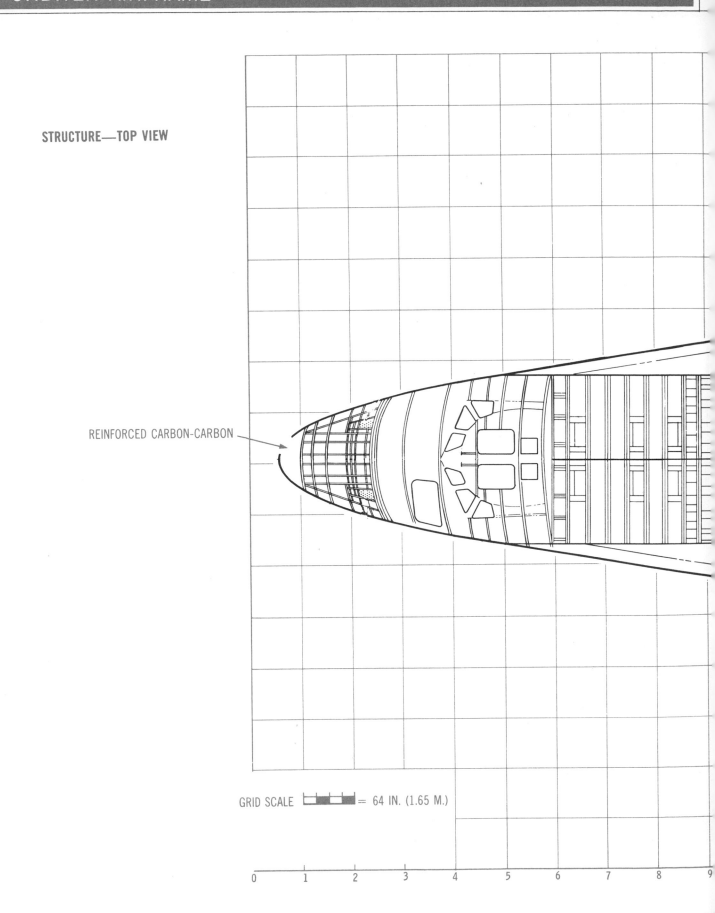

STRUCTURE—TOP VIEW

REINFORCED CARBON-CARBON

GRID SCALE = 64 IN. (1.65 M.)

0   1   2   3   4   5   6   7   8   9

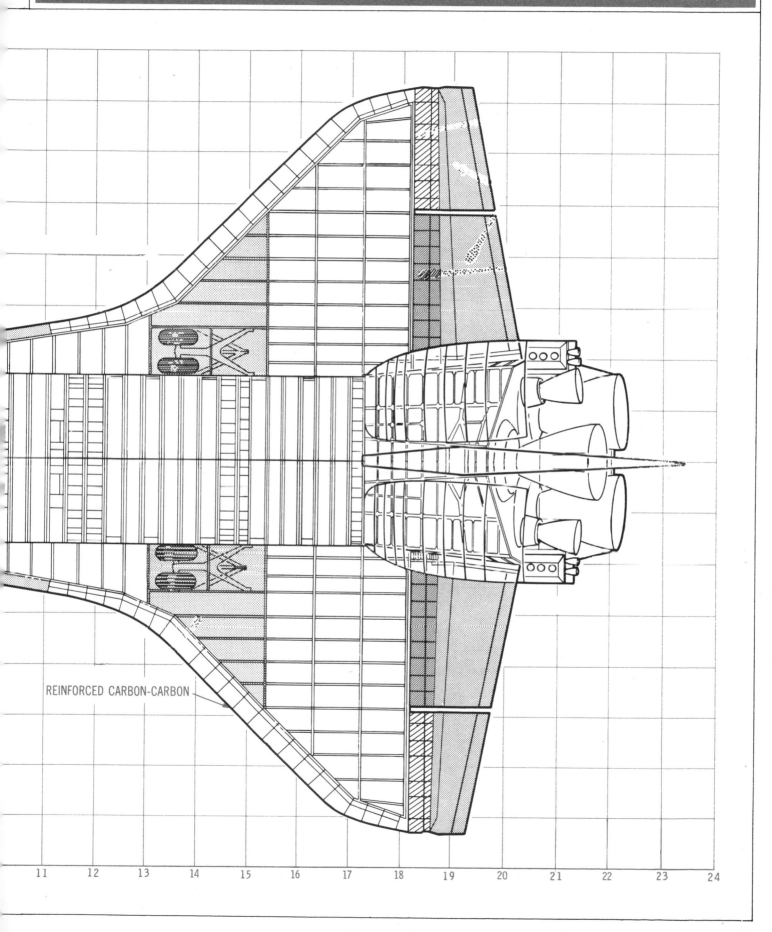

REINFORCED CARBON-CARBON

STRUCTURE—BOTTOM VIEW

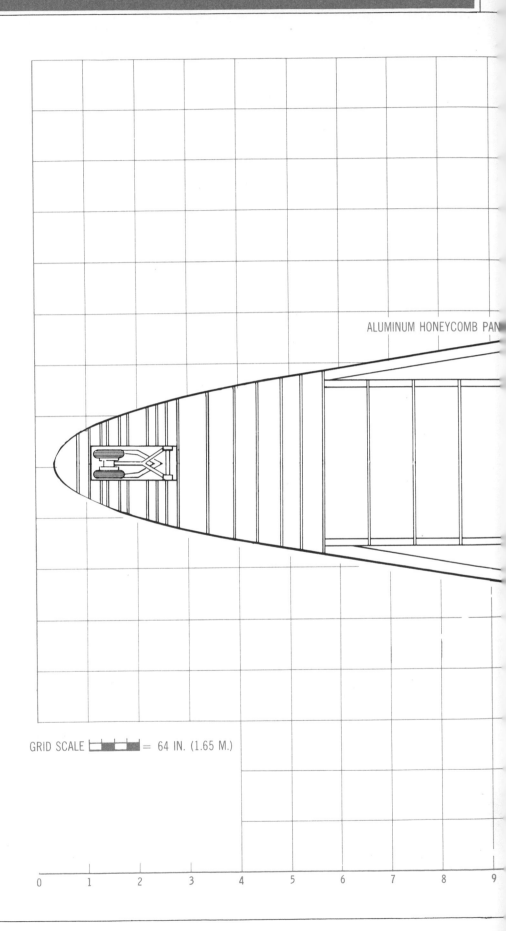

ALUMINUM HONEYCOMB PAN

GRID SCALE ⊏▭⊐ = 64 IN. (1.65 M.)

0   1   2   3   4   5   6   7   8   9

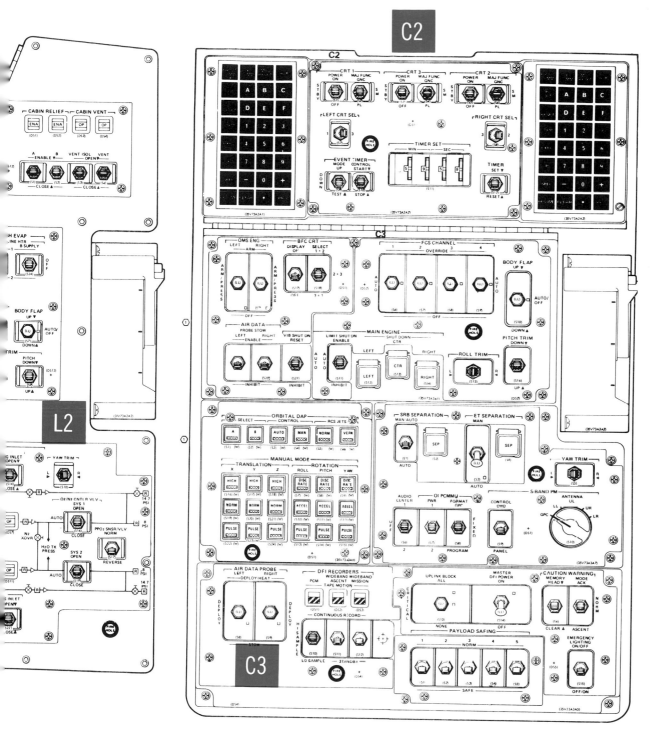

FLIGHT DECK CONSOLES
L1, L2, C2, C3, R1, and R2

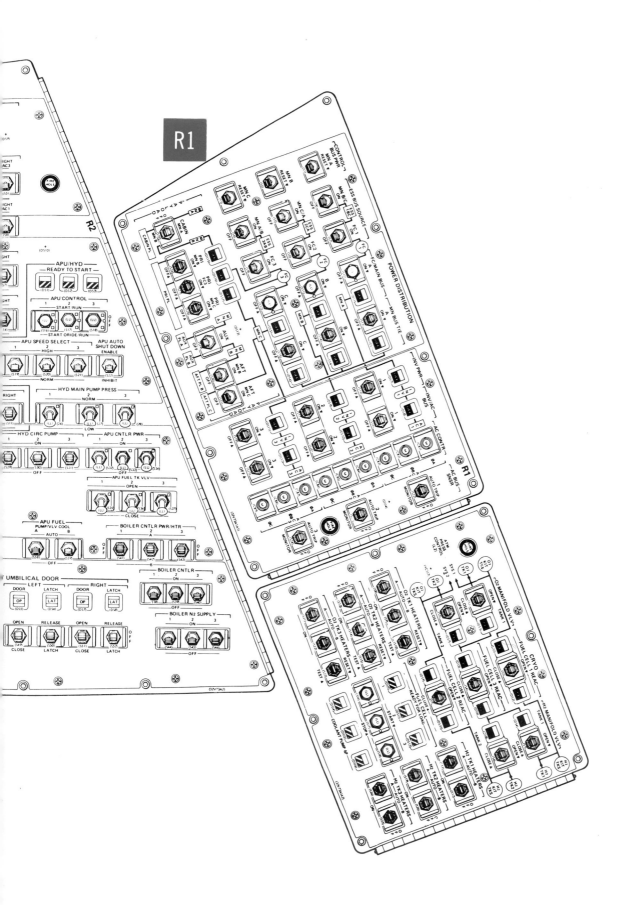

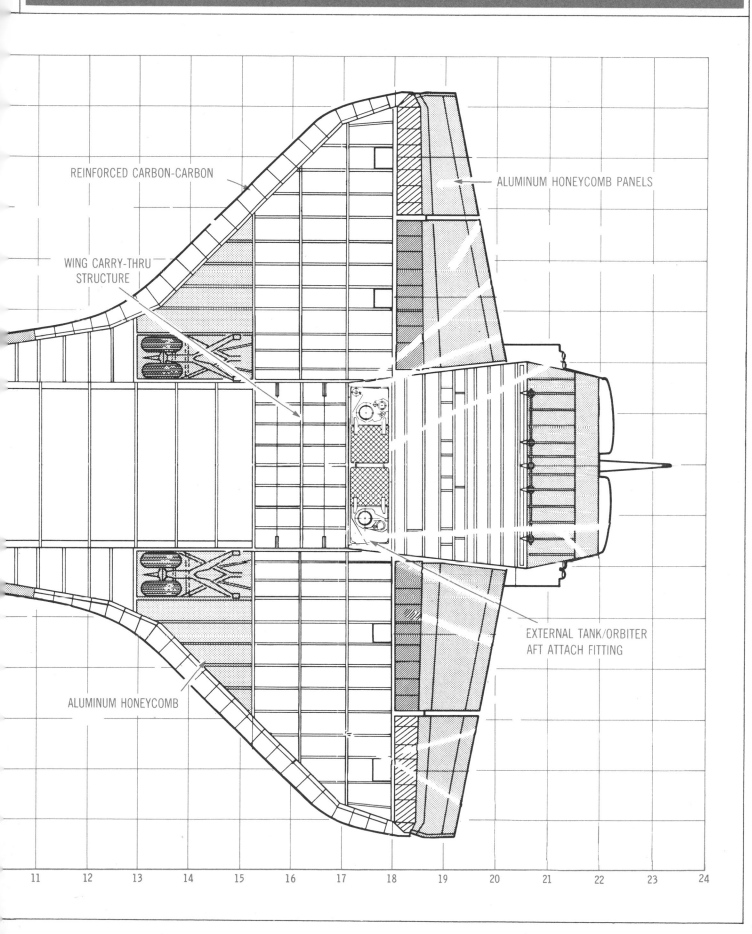

REINFORCED CARBON-CARBON

ALUMINUM HONEYCOMB PANELS

WING CARRY-THRU
STRUCTURE

EXTERNAL TANK/ORBITER
AFT ATTACH FITTING

ALUMINUM HONEYCOMB

11 12 13 14 15 16 17 18 19 20 21 22 23 24

BODY PENETRATIONS/ACCESS PANELS, LEFT SIDE

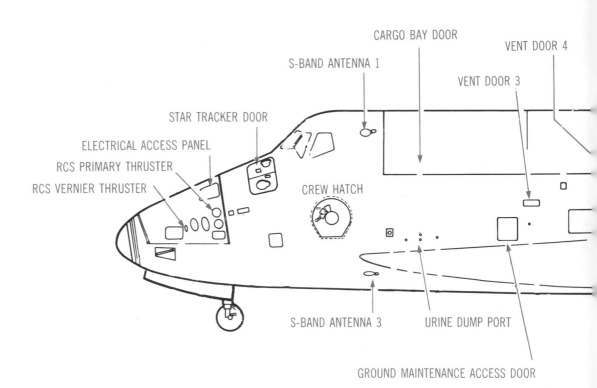

CARGO BAY DOOR

VENT DOOR 4

S-BAND ANTENNA 1

VENT DOOR 3

STAR TRACKER DOOR

ELECTRICAL ACCESS PANEL

RCS PRIMARY THRUSTER

RCS VERNIER THRUSTER

CREW HATCH

S-BAND ANTENNA 3

URINE DUMP PORT

GROUND MAINTENANCE ACCESS DOOR

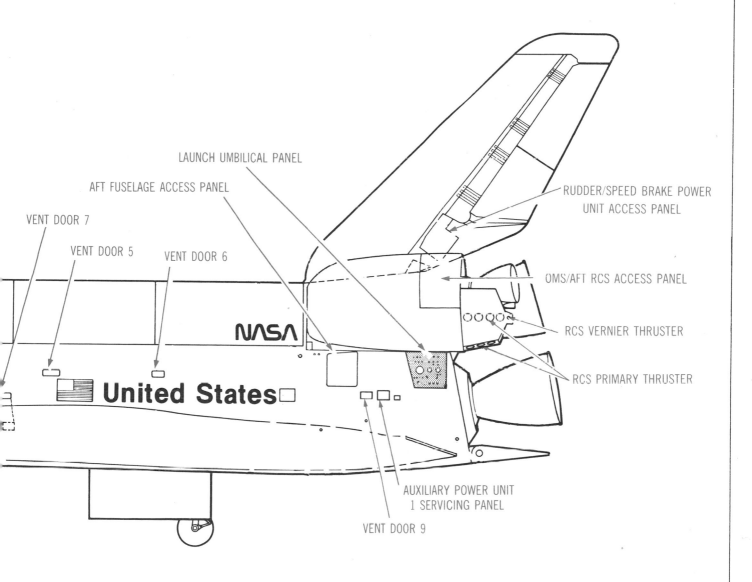

LAUNCH UMBILICAL PANEL

AFT FUSELAGE ACCESS PANEL

VENT DOOR 7

VENT DOOR 5   VENT DOOR 6

NASA

United States

RUDDER/SPEED BRAKE POWER
UNIT ACCESS PANEL

OMS/AFT RCS ACCESS PANEL

RCS VERNIER THRUSTER

RCS PRIMARY THRUSTER

AUXILIARY POWER UNIT
1 SERVICING PANEL

VENT DOOR 9

BODY PENETRATIONS/ACCESS PANELS, TOP

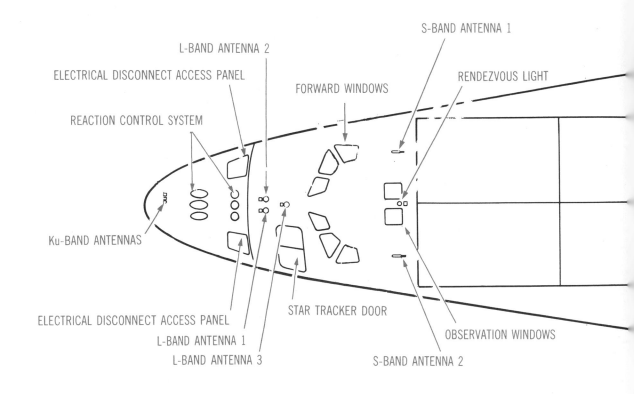

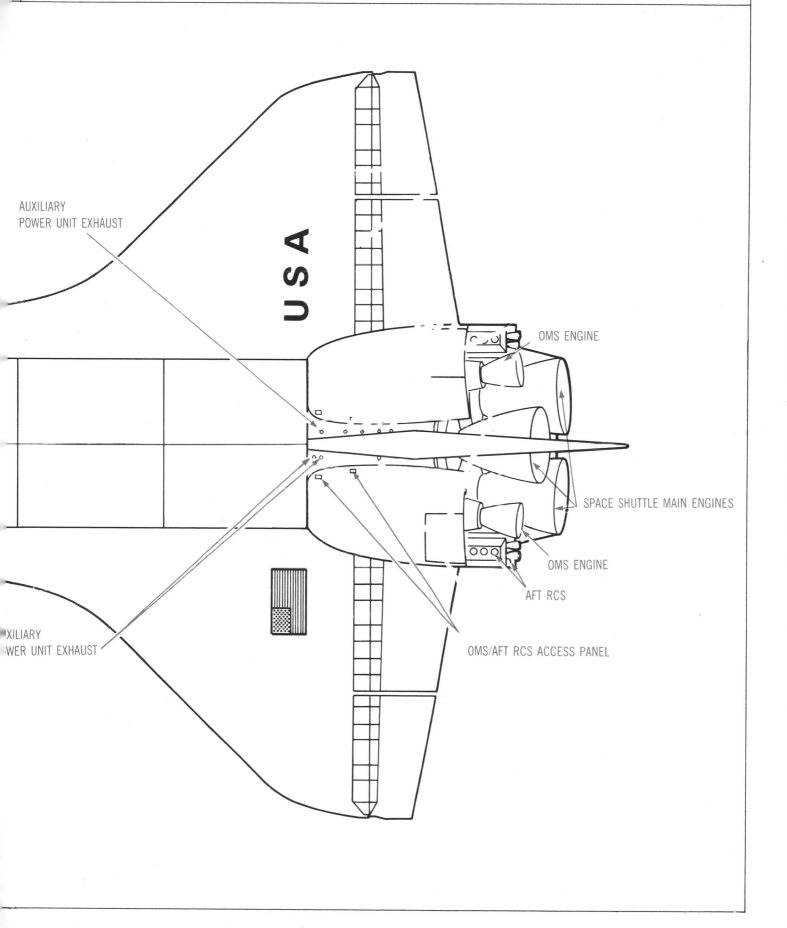

AUXILIARY
POWER UNIT EXHAUST

USA

OMS ENGINE

SPACE SHUTTLE MAIN ENGINES

OMS ENGINE

AFT RCS

OMS/AFT RCS ACCESS PANEL

XILIARY
WER UNIT EXHAUST

BODY PENETRATIONS/ACCESS PANELS, BOTTOM

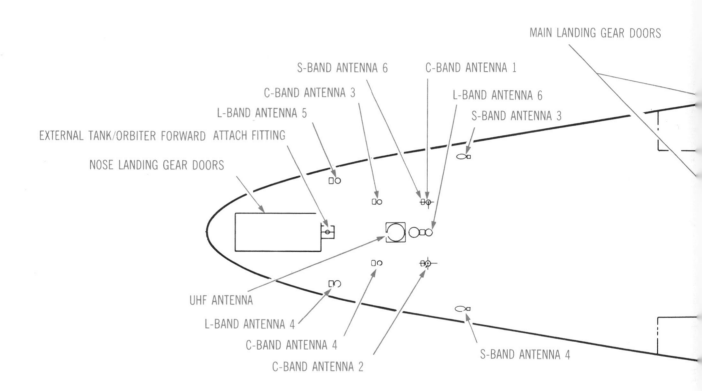

MAIN LANDING GEAR DOORS

S-BAND ANTENNA 6

C-BAND ANTENNA 1

C-BAND ANTENNA 3

L-BAND ANTENNA 6

L-BAND ANTENNA 5

S-BAND ANTENNA 3

EXTERNAL TANK/ORBITER FORWARD ATTACH FITTING

NOSE LANDING GEAR DOORS

UHF ANTENNA

L-BAND ANTENNA 4

C-BAND ANTENNA 4

C-BAND ANTENNA 2

S-BAND ANTENNA 4

NOTE: ALL ANTENNAS ARE COVERED WITH THERMAL PROTECTION INSULATION

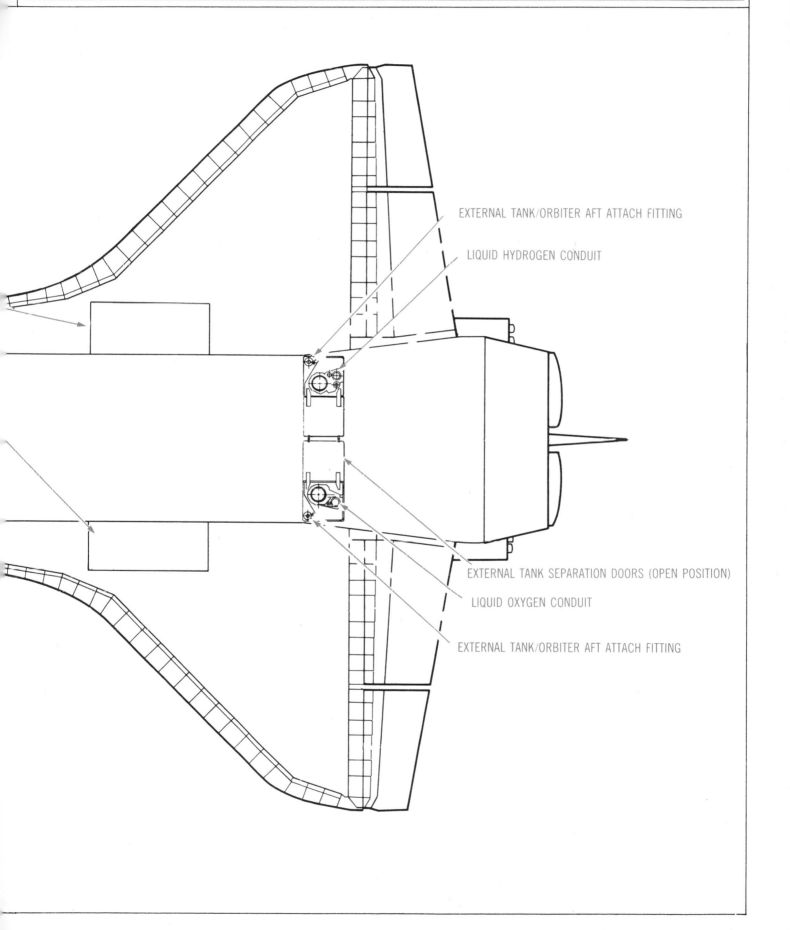

EXTERNAL TANK/ORBITER AFT ATTACH FITTING

LIQUID HYDROGEN CONDUIT

EXTERNAL TANK SEPARATION DOORS (OPEN POSITION)

LIQUID OXYGEN CONDUIT

EXTERNAL TANK/ORBITER AFT ATTACH FITTING

Thermal Protection System

Ablative heatshields protected Mercury, Gemini, and Apollo spacecraft from the searing heat of entry. During entry, such heatshields char and vaporize. This action carries excess heat away from the spacecraft. However, ablative heatshields have two disadvantages: they are heavy, and they can be used only once.

The reusable Space Shuttle has a new type of lightweight reusable surface insulation. Called the **thermal protection system**, or TPS, it keeps the Orbiter structure below 175°C (350°F) during entry. Several different materials make up the TPS. A high-temperature, high-strength material called **reinforced carbon-carbon** (RCC) is used on the nosecap and leading edges of the wings. This gray substance can withstand temperatures of 1,650°C (3.000°F).

The black areas on the Orbiter are covered with blocks or tiles of varying size and thickness. These are the **high-temperature reusable surface insulation**, or HRSI. HRSI tiles protect areas where temperatures are between 650°C (1,200°F) and 1,275°C (2,300°F).

On Orbiter 102, **Columbia**, white tiles cover the forward fuselage, outer wing areas, pods, and the stabilizer. Called **low-temperature reusable surface insulation**, or LRSI, these tiles are used where temperatures are between 370°C (700°F) and 650°C (1,200°F).

Columbia's cargo-bay doors, fuselage sides, upper wing surfaces, and aft areas of the OMS pods are covered with a **Nomex felt** material. These areas remain below 370°C (700°F) during flight.

Altogether, nearly 32,000 HRSI and LRSI tiles cover **Columbia**. No two tiles are alike and each must be installed by hand. Both types of tiles are made from extremely pure (99.7%) **sand**! The sand is crushed into very small silica fibers and added to a ceramic binder. This mixture is fired to produce the blocks. They are mchined to the proper size and shape, then the black or white coating is applied to their outer surfaces. The coating is made from a high-strength refractory glass.

An aluminum structure like that of the Orbiter flexes and bends slightly in flight. The TPS tiles covering the vehicle must be very close together. On the underside, the largest allowable gap between tiles is only 0.065 inch (1.6 millimeters). These glass-covered silica tiles are rather brittle and cannot flex or bend without breaking. To let the structure flex while keeping the TPS rigid, Nomex felt pads are sandwiched between the tiles and the structure. This way, the structure can move without moving the tiles. The pads and tiles are attached with a thin layer of a room temperature vulcanizing silicon adhesive.

HRSI tiles are usually 6 by 6 inches (15 by 15 centimeters); LRSI tiles are 8 by 8 inches (20 by 20 centimeters). However, in areas with compound curves, the sizes and shapes of the tiles varies. In general, HRSI tiles are thicker than the LRSI.

On Orbiter 099, **Challenger**, some of the LRSI tiles have been replaced by an advanced fabric insulation. This insulation is a quilted fabric similar to the material used on **Columbia** only thicker. This thickened fabric insulation has replaced all the LRSI tiles on Orbiters 103 and 104.

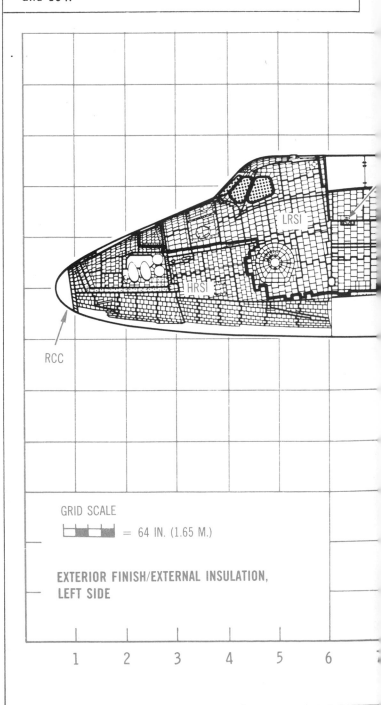

RCC

GRID SCALE

= 64 IN. (1.65 M.)

EXTERIOR FINISH/EXTERNAL INSULATION, LEFT SIDE

1 2 3 4 5 6

NOTE

RCC—REINFORCED CARBON-CARBON
  [0.25 IN. (0.63 cm) TO 0.5 IN. (1.27 cm)]

HRSI—HIGH-TEMPERATURE REUSABLE SURFACE INSULATION
  [0.5 IN. (1.27 cm) TO 5 IN. (12.7 cm)]

LRSI—LOW-TEMPERATURE REUSABLE SURFACE INSULATION
  [0.20 IN. (0.51 cm) TO 2.75 IN. (6.98 cm)]

FRSI—REUSABLE NOMEX FELT SURFACE INSULATION
  [0.144 IN. (0.36 cm) TO 1.1 IN. (2.79 cm)]

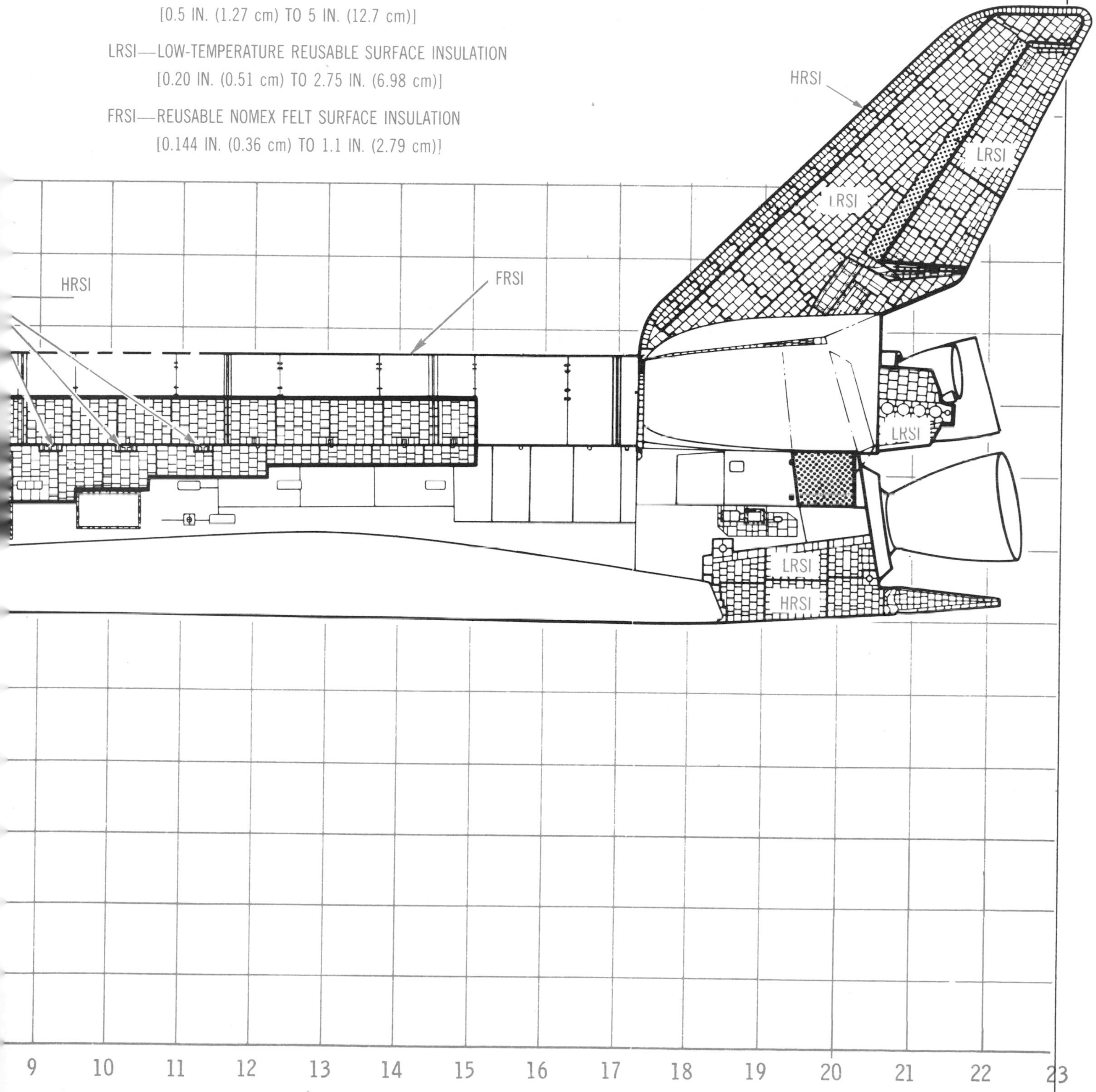

HRSI

FRSI

HRSI

LRSI

LRSI

LRSI

LRSI

HRSI

9 10 11 12 13 14 15 16 17 18 19 20 21 22 23

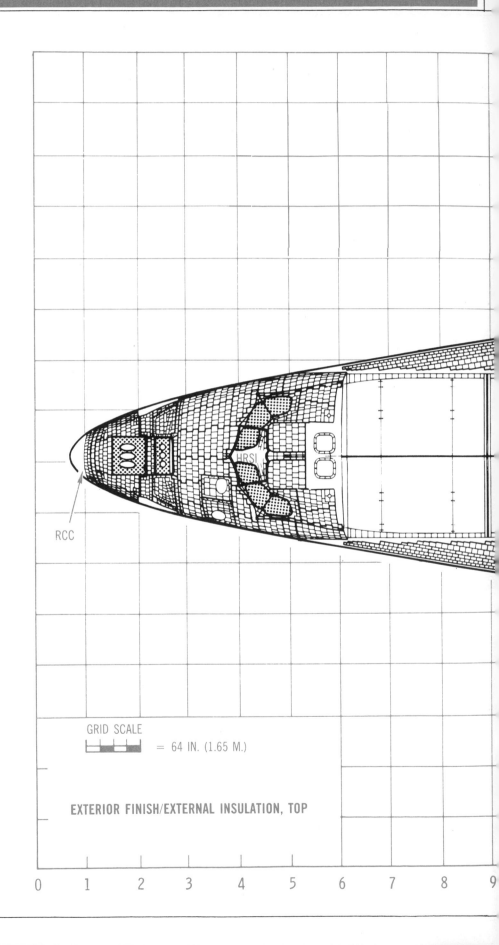

RCC

HRSI

GRID SCALE

= 64 IN. (1.65 M.)

EXTERIOR FINISH/EXTERNAL INSULATION, TOP

0 1 2 3 4 5 6 7 8 9

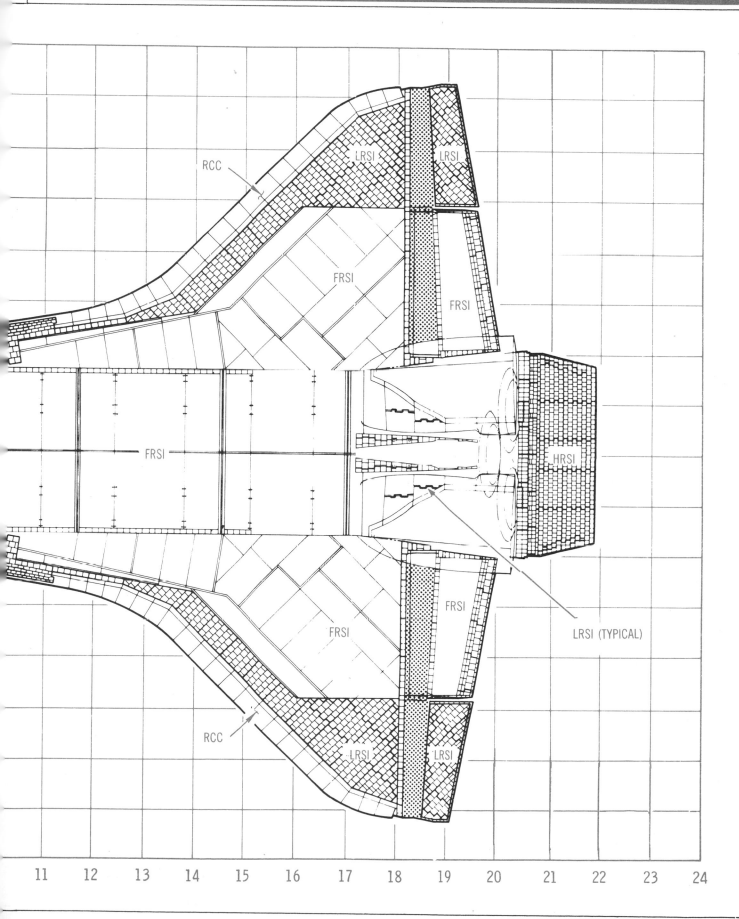

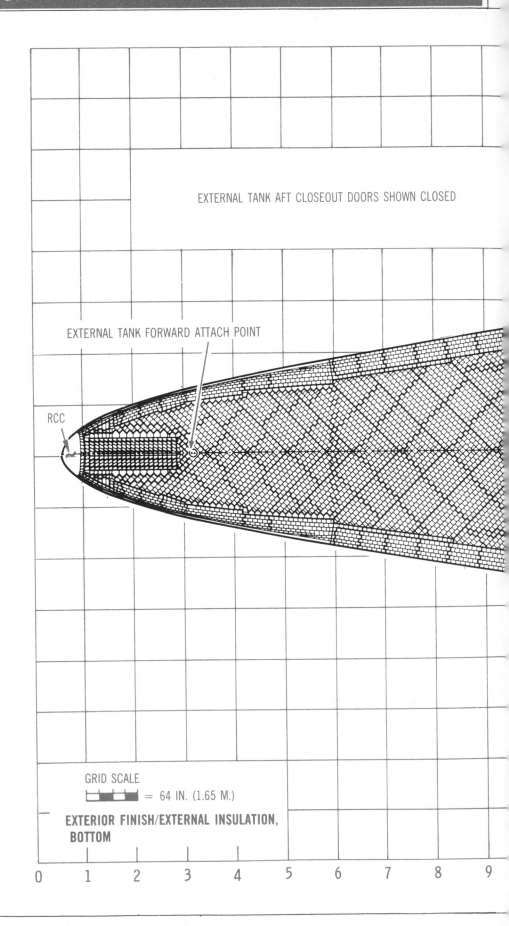

EXTERNAL TANK AFT CLOSEOUT DOORS SHOWN CLOSED

EXTERNAL TANK FORWARD ATTACH POINT

RCC

GRID SCALE

= 64 IN. (1.65 M.)

EXTERIOR FINISH/EXTERNAL INSULATION, BOTTOM

0   1   2   3   4   5   6   7   8   9

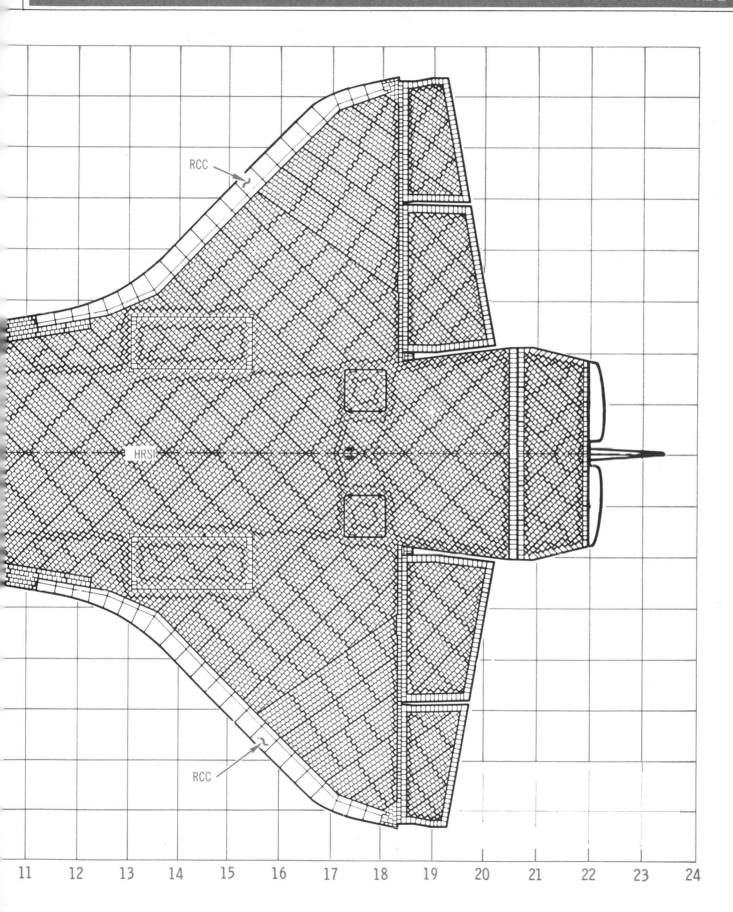

Space Shuttle Main Engine

The **space shuttle main engine** (SSME) is one of the most advanced rocket engines ever built. It burns liquid hydrogen and liquid oxygen to produce a rated thrust of 375,000 pounds (1,700,000 newtons) at sea level. The thrust can be varied from 65% to 109% of the rated value.

Using the propellants to power the turbopumps results in a very efficient engine. Further, feeding the propellants at high pressures to the combustion chamber gives the highest possible thrust for the engine's size.

There are three SSMEs on the Orbiter's aft end. Each engine was designed to operate for a total of up to 7½ hours between major overhauls. At a rate of only 8 minutes per flight, the engines should last for fifty-five missions.

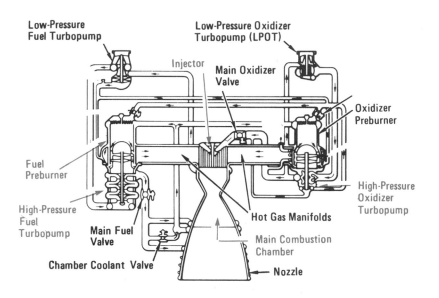

During operation, the hydrogen fuel is fed to a pair of **preburners**. On the way to the preburners, some super-cold liquid hydrogen is diverted from the mainstream and circulated through tubes in the engine's nozzle and combustion chamber walls. This cools the engine. Eventually, however, all the hydrogen finds its way to the preburners. The mixture ratio of oxygen to hydrogen is less than one part oxygen to one part hydrogen. A very hot hydrogen-rich steam results from this mixture. The hot gas drives the high-pressure propellant **turbopumps** before it is ducted into the main **injector**, where it mixes with the rest of the oxygen and is finally fed to the **combustion chamber**. The mixture ratio of oxygen to hydrogen for final combustion in 6:1.

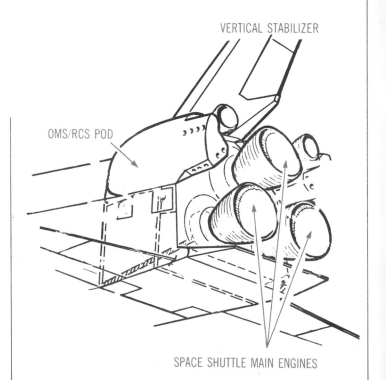

SPACE SHUTTLE MAIN ENGINES

Orbital Maneuvering System

Two **orbital maneuvering system** (OMS) engines, mounted in external pods on each side of the aft fuselage, power the Orbiter during orbital insertion and deorbit. Aditionally, the OMS provides thrust for large orbital changes.

Each engine has a thrust of 6,000 pounds (26,700 newtons). The propellants are monomethyl hydrazine (the fuel) and nitrogen tetroxide (the oxidizer). Helium gas forces the propellants from their tanks and into the engines. Propellants for each engine are contained in their respective pods. However, there is a cross-feed system to transfer propellants from one pod to the other if needed.

The OMS engine is designed to last a hundred missions. It is 77 inches (2 meters) long and weighs 260 pounds (118 kilograms). The engine is gimbaled in pitch and yaw.

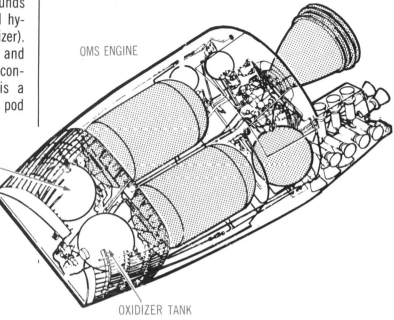

OMS ENGINE

FUEL TANK

OXIDIZER TANK

Reaction Control System

The **reaction control system** (RCS) is a system of forty-four small rocket engines that maneuver the Orbiter in space. There are thirty-eight primary RCS thrusters, each with 870 pounds thrust (3,870 newtons), and six 25-pound-thrust (110-newton) vernier engines.

The RCS thrusters are grouped in three modules, one in the Orbiter nose and one in each OMS pod. They use the same propellants as the OMS engies. In the aft pods, the RCS has its own propellant tanks. However, propellants can be fed to the RCS tanks from the OMS tanks.

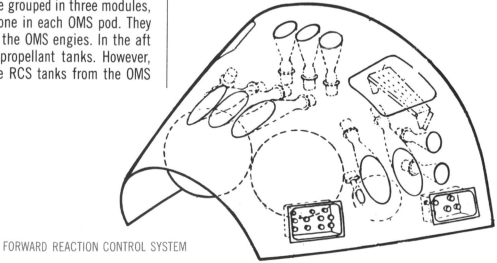

FORWARD REACTION CONTROL SYSTEM

Auxiliary Power Unit

Three **auxiliary power units** (APUs) generate power for the Orbiter's hydraulic system. This system is very important—it provides power to move the SSMEs during ascent and to adjust the Orbiter's control surfaces during descent.

Two APUs are sufficient, but a third is provided as a backup. As additional backup insurance, there are three separate and independent hydraulic systems.

The APU uses hydrazine for its fuel. Each unit has its own fuel supply—295 pounds (134 kilograms) of hydrazine, enough for 90 minutes. The APUs are turned on only during ascent and return.

Each APU weighs about 88 pounds (40 kilograms) and generates 135 horsepower.

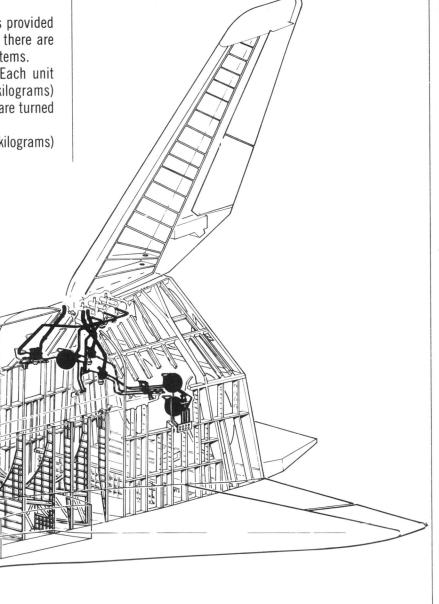

Landing Gear

The Orbiter has three **landing-gear** assemblies—two main ones and one in the nose. Each has two tires. When you push the DN pushbutton to extend the landing gear, the unlock hooks are released via hydraulic power, then the assemblies fall to the down position, assisted by springs and hydraulic power. When fully deployed, the gear locks in position.

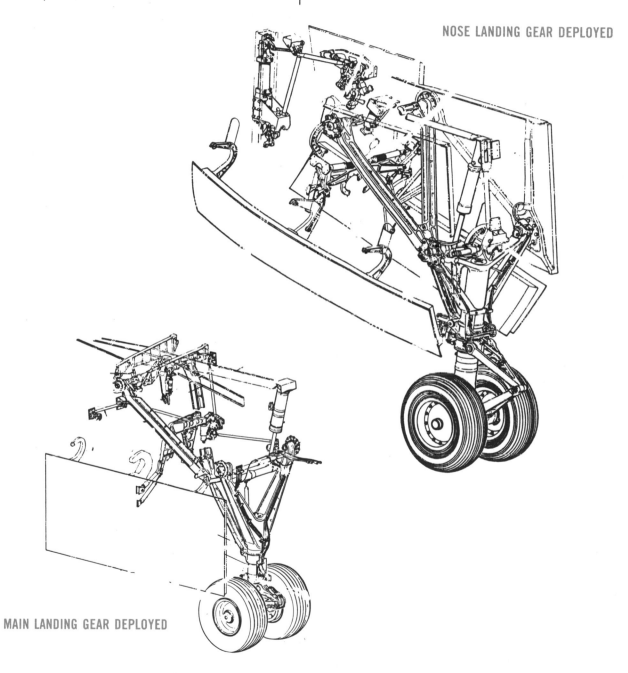

NOSE LANDING GEAR DEPLOYED

MAIN LANDING GEAR DEPLOYED

ELECTRICAL POWER SYSTEM

THREE HYDROGEN-OXYGEN FUEL CELLS GENERATE
28 D.C. VOLTS WITH A NORMAL OUTPUT OF
1000-1500 WATTS FOR ORBITER USE. FOR PAYLOAD
OPERATION, AN AVERAGE OF 7000 WATTS ARE AVAILABLE.

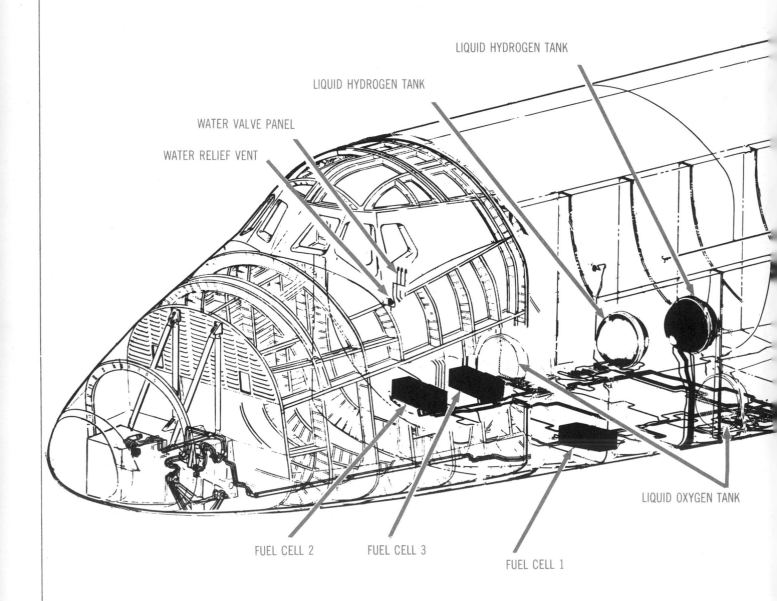

LIQUID HYDROGEN TANK

LIQUID HYDROGEN TANK

WATER VALVE PANEL

WATER RELIEF VENT

LIQUID OXYGEN TANK

FUEL CELL 2

FUEL CELL 3

FUEL CELL 1

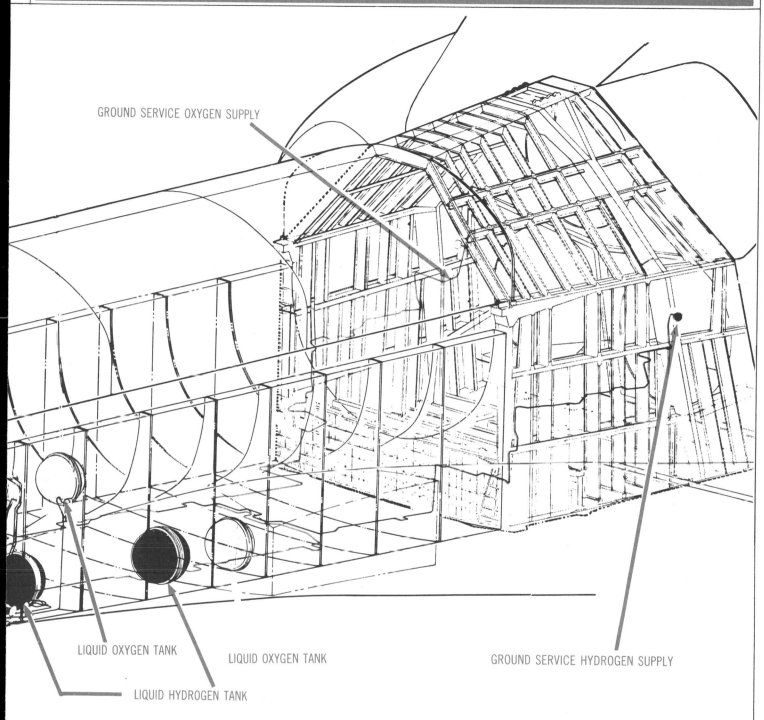

GROUND SERVICE OXYGEN SUPPLY

LIQUID OXYGEN TANK

LIQUID OXYGEN TANK

GROUND SERVICE HYDROGEN SUPPLY

LIQUID HYDROGEN TANK

External Tank

The External Tank (ET), the largest single piece of the Space Shuttle, is the only major component not reused. It holds liquid oxygen and liquid hydrogen for the trio of main engines on the Orbiter's base. About 8½ minutes after launch, the propellants in the tank are consumed and it is discarded. It breaks apart in the upper atmosphere over a remote ocean area.

Overall Dimensions

Length:	154 feet (47 meters)
Diameter:	27.5 feet (8.4 meters)
Weight (empty):	78,100 pounds (35,400 kilograms)

Structure

The ET comprises two separate tanks connected by a corrugated intertank section. The liquid-oxygen tank is the smaller **forward tank**; liquid hydrogen is contained in the **aft tank**.

Aluminum alloys are used for the ET. Its exterior is covered with materials which protect it from extreme temperatures during launch and ascent. A spray-on polyurethane foam covers and insulates the entire tank. Over that, an ablative material protects the tank's nose and base, the areas that get hottest.

The first two ETs flown were painted white. On subsequent flights, the tanks were left in the beige color of the insulation to reduce their weight.

Volume and Propellant Weights

Liquid Oxygen Tank:	143,000 gallons (540,000 liters), 1,359,000 pounds (616,500 kilograms)
Liquid Hydrogen Tank:	383,000 gallons (1,500,000 liters), 226,000 pounds (102,000 kilograms)

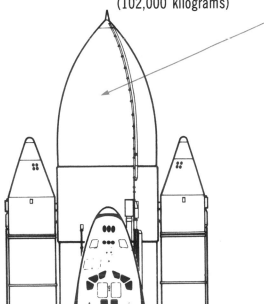

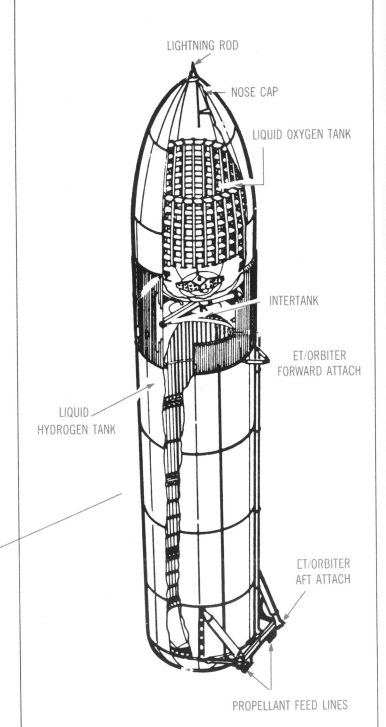

LIGHTNING ROD

NOSE CAP

LIQUID OXYGEN TANK

INTERTANK

ET/ORBITER FORWARD ATTACH

LIQUID HYDROGEN TANK

ET/ORBITER AFT ATTACH

PROPELLANT FEED LINES

Solid Rocket Boosters

Two large solid-propellant rockets operate in parallel with the main engines during the first 2 minutes of flight. The Solid Rocket Boosters (SRBs) provide most of the power needed to lift the Shuttle during this part of the flight. The SRB is the first solid rocket motor designed for reuse and the largest solid motor ever flown.

The SRB **nozzles** swivel (or **gimbal**) up o 6° to direct the thrust and steer the Shuttle.

Overall Dimensions (Each Booster)

Length: 149 feet (45.5 meters)
Diameter: 12 feet (3.6 meters)
Weight at launch: 1,300,000 pounds (590,000 kilograms)
Thrust at launch: 2,650,000 pounds (11,800,000 newtons)

Propellants

Fuel: Aluminum powder, 16%
Oxidizer: Ammonium perchlorate, 69.83%
Catalyst: Iron oxide powder, 0.17%
Other ingredients: Binder and curing agent, 14%

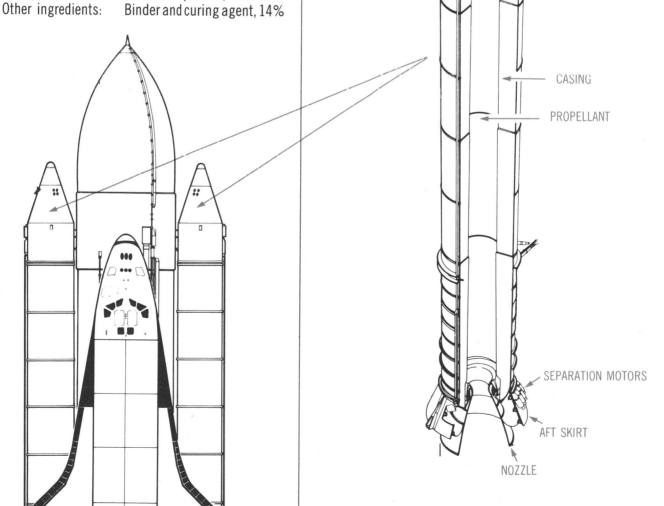

DROGUE PARACHUTE

SEPARATION MOTORS

MAIN PARACHUTES (3)

CASING

PROPELLANT

SEPARATION MOTORS

AFT SKIRT

NOZZLE

Kennedy Space Center

The primary launch and landing site for the Space Shuttle is the Kennedy Space Center at Cape Canaveral, Florida, the original launch site of all U.S. manned missions. The lunar launching pads of Launch Complex 39 have been modified to serve as Shuttle launch pads. The **Vehicle Assembly Building** (VAB) where the thirty-six-story Apollo-Saturn components were erected is still used, only now it is used to put the Orbiter, External Tank, and Solid Rocket Boosters together. But there are important changes.

Behind the VAB is the **Orbiter Processing Facility** (OPF). After launching, the Orbiter is rolled into this facility. There are two huge bays almost 200 feet (61 meters) long, 150 feet (46 meters) wide, and 95 feet (29 meters) high. Here two Orbiters can be thoroughly serviced and checked out and can have horizontal payloads like Spacelab removed and unloaded. A service structure allows access to all parts of the Orbiter, including the thermal protection tiles.

From the Orbiter Processing Facility, the Orbiter is rolled to the Vehicle Assembly Building, which covers 8 acres (3.25 hectares) and stands 525 feet (160 meters) tall. Here Shuttles are erected on a mobile launch pad. Huge cranes lift the Orbiter and join it with the SRBs and External Tank. Here also the system is checked out prior to rollout to the launch pad.

Attached to the VAB is the **Launch Control Center** containing the two Shuttle firing rooms. About forty-five people are required for checkout and launch operations. This is only a tenth as many as were required for an Apollo launch.

The **Mobile Launch Platform** and the Spacecraft are moved to the launch pad by one of the two crawler transporters, the largest land vehicles in the world. Each 3,000 ton (2,700,000 kilogram) vehicle moves on four double-tracked crawlers at a maximum speed of 2 mph (3.2 kph).

The launch pads themselves are complex structures. A fixed **Service Structure** gives access to all necessary Orbiter parts with extending arms. A **Rotating Service Structure** can be rolled to cover the Orbiter. Here vertical payloads like satellites can be loaded and serviced on the pad. Surrounding the pad are the large fuel tanks from which the tanks on the Shuttle are filled. Under the pad is a flame trench with a sound- and shockwave suppressive system.

The landing facilities include a 15,000-foot (4,600-meter)-long, 300-foot (91-meter)-wide runway. This is 50% longer than standard runways and has a 1,000-foot (300-meter) overrun at each end for added safety. The extra length is necessary because with a glide landing there is only one chance to get it right.

A microwave landing system, part of the landing facilities, supplies accurate data to the Orbiter to ensure a safe landing. Distance and elevation measurements are continuously computed and provided to the Orbiter's decoder-receivers.

At the end of this runway is a mate-demate device for loading and unloading the Orbiter from the 747 ferry aircraft that can carry the Orbiter between launch sites when required.

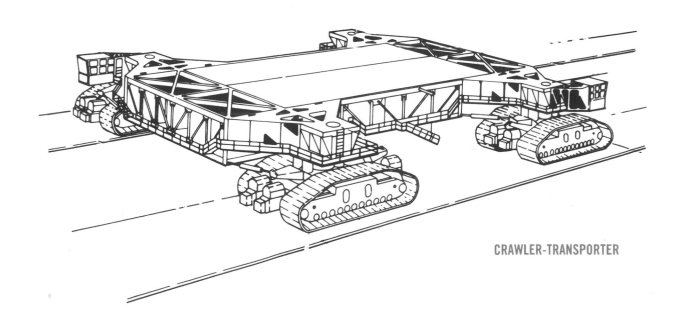

CRAWLER-TRANSPORTER

Vandenberg Air Force Base and Western Test Range

Launches from the Kennedy Space Center head eastward. This puts the Shuttle in roughly equatorial orbits. These are the easiest orbits to reach, because the spin of the earth—about 1,000 mph (1,600 kph)—gives an extra push. However, many payloads require orbits over the poles. Observations for meteorology, remote earth sensing, and the like use polar orbits. To service and deploy such satellites, shuttles are launched southward down the Western Test Range from Vandenberg Air Force Base in Lompoc, California.

The facilities at Vandenberg parallel but do not duplicate those at the Kennedy Space Center. A more modest Orbiter processing facility permits servicing and payload integration. Erection cranes perform the same work as those in the Vehicle Assembly Building, but it is done outdoors, and the Orbiter, tank, and solid boosters slide into position over the pad. A vibration and shockwave reduction system is also built into the pad.

At the pad, a mobile service tower is available. A payload changeout room, and the necessary fuel tanks are provided.

A solid Rocket Booster refurbishment system completes this launch complex.

Landing Systems and Facilities

During descent and landing, three major radio-electronic systems provide you with precise position-indication information. TACAN, which is available after you leave the blackout, gives you both range and bearing measurements. At 18,000 feet (5,500 meters), the microwave landing system can be used. This system gives you the angle between your course and the desired path or trajectory, and your distance from the runway. Below 9,000 feet (2,750 meters), a radar altimeter is added.

Two Airdata probes provide additional information about altitude.

TACAN, short for **tactical air navigation**, is comprised of a series of ground stations. Their transmissions at specific frequencies are detected by the Orbiter. From these, the distance from and bearing to the station can be determined. This is updated every 37 seconds. The on-board TACAN actually computes the angle between lines from the spacecraft to magnetic north and to the ground station being detected; this provides bearing. The on-board unit also signals the ground station, and the length of time it takes to reply supplies the distance information.

The **microwave landing system** takes over when you approach the heading alignment cylinder. This system provides the angle of **elevation** (up and down), the angle of **azimuth** (left and right), and your range. The beam scans about 15° to the right or left of the runway centerline and about 30° of vertical coverage. As you deviate from your "ideal" flight path, you would see if you needed to correct left, right, up, or down to stay on that path.

Precise altitude information is provided by your two onboard **radar altimeters**. The difference in time for an emitted pulse to return provides your height above the ground. This is critical during touchdown because it provides precise data for the sink rate and the height of the rear wheels above the runway.

All of this equipment is available at your primary and secondary landing sites. The primary landing sites are at the Kennedy Space Center and Vandenberg Air Force Base. Secondary sites are at the Dryden Flight Research Center.

The major abort landing facilities are Kennedy Space Center; Rota, Spain; and Northrup Strip, New Mexico. Landing strips in Okinawa and Hawaii can be used in extreme emergencies.

Control Centers

Throughout your flight, you speak to **flight controllers** at various **control centers**. These centers provide necessary engineering and operational data to the crew. In an emergency, the centers have access to the necessary experts to solve problems. These experts often have identical equipment on the ground to simulate both normal operations and problems. On even the best-planned missions, technical questions will arise. You can rely on the control centers for accurate instructions and for attention to mission safety.

The Launch Control Center is responsible for spacecraft and pad checkout, prelaunch countdown, and all launch operations until you are about 10 seconds into your flight.

Once you have cleared the tower, you are handed over to the Mission Control Center. Three teams operate this center on three around-the-clock shifts during the mission. Data, voice, and video communications from the Orbiter are relayed to Mission Control. Automated commands for the Orbiter and data processing also originate here. A **capsule communicator** (CAP COM), usually one of the other astronauts, maintains contact with the Orbiter crew.

Special payload operations are conducted by Payload Operation Control Centers. Here, experts have special communication and display equipment to control payloads and advise the crew.

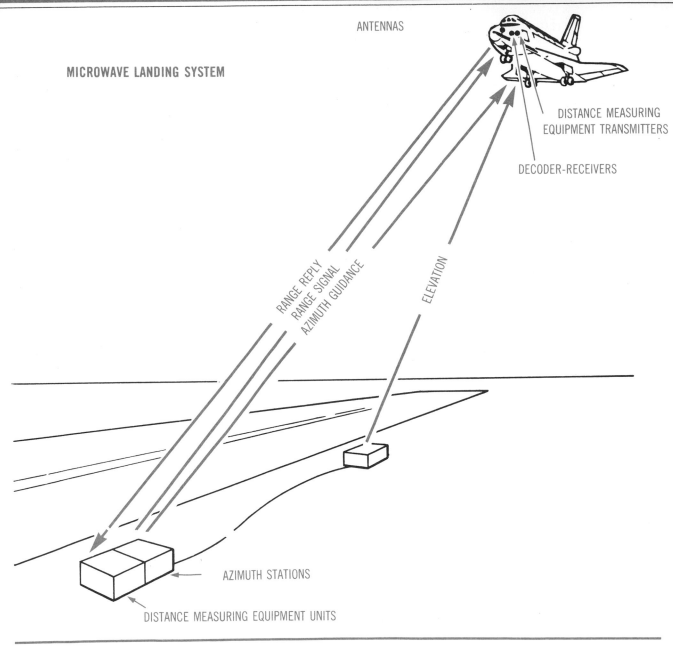

MICROWAVE LANDING SYSTEM

ANTENNAS

DISTANCE MEASURING
EQUIPMENT TRANSMITTERS

DECODER-RECEIVERS

RANGE REPLY
RANGE SIGNAL
AZIMUTH GUIDANCE

ELEVATION

AZIMUTH STATIONS

DISTANCE MEASURING EQUIPMENT UNITS

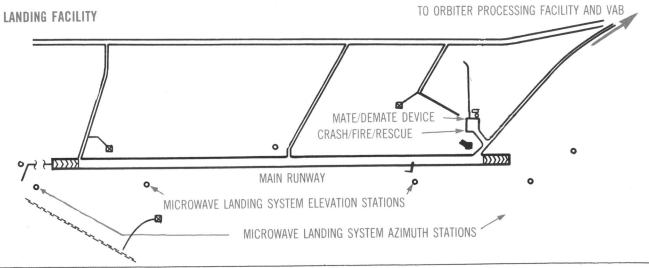

LANDING FACILITY

TO ORBITER PROCESSING FACILITY AND VAB

MATE/DEMATE DEVICE
CRASH/FIRE/RESCUE

MAIN RUNWAY

MICROWAVE LANDING SYSTEM ELEVATION STATIONS

MICROWAVE LANDING SYSTEM AZIMUTH STATIONS

Shuttle Carrier Aircraft

For cross-country movement, the Orbiter rides piggy-back atop a Boeing 747-100 jumbo jet. This mode of transport is used to ferry the craft to and from various support facilities.

After NASA purchased the 747 from American Airlines in 1974, they outfitted it with more powerful engines and modified it to accommodate the 75-ton (68,000 kilogram) Orbiter. All passenger equipment—seats, galleys, etc.—was removed. Additional bulkheads and structural supports were added inside the aircraft. Three supports for the Shuttle were placed on the jet's top, and tip fins were added to the horizontal stabilizer.

A tapered tail fairing fits over the Orbiter's aft end during transport. The fairing smooths the airflow around the Orbiter, reducing aerodynamic drag while it is in flight.

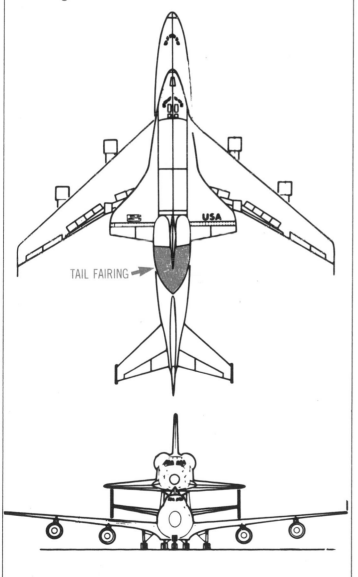

TAIL FAIRING →

Solid Rocket Booster Recycling

The Shuttle's Solid Rocket Boosters (SRBs) are the first solid-propellant rocket motors designed for reuse.

During the first 2 minutes of flight, two SRBs provide most of the power needed for the Shuttle's ascent. After they burn out, four small motors on each end of the SRB casings fire, pushing the cases away from the External Tank and Orbiter. The nozzle extensions separate 70 seconds later, as the casings fall to earth. After 2 more minutes, the nosecaps are jettisoned. Small drogue parachutes pop out, followed by three 115-foot (35-meter)-diameter main parachutes on each booster. The boosters land base-first in the ocean.

Two ocean-going tugs locate the boosters with tracking and sonar beacons. Each tug recovers one booster. Air in the casing keeps it floating upright. After the casing is "safed," a towline is attached. A remote-controlled nozzle plug seals the open end of the casing. When the water is pumped out of the casing, it floats horizontally like a log, and is then towed to port.

In port, the casing is broken down into four sections. The sections are washed clean of any remaining insulation and propellant, using water jets with a pressure of 6,000 psi (41,000 kilopascals). These four pieces are then broken down into the original eleven segments, which are degreased and grit-blasted. The segments are then tested for cracks. Then they are pressure-tested with oil and once more checked for cracks. The segments are then put back together and repainted, and are then ready to be filled and reused. It is hoped that these will last for about twenty missions, but that depends upon test results after each flight.

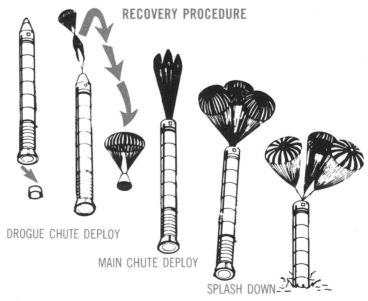

RECOVERY PROCEDURE

DROGUE CHUTE DEPLOY

MAIN CHUTE DEPLOY

SPLASH DOWN

abort —To end the mission short of its objective. An abort is usually caused by some malfunction or emergency.

accelerometer —An instrument that measures the rate at which acceleration changes.

airlock —A chamber used to adjust pressure for passage between one area and another; for example, between the Orbiter and the outside space environment.

apogee —The highest point of an earth ORBIT.

apogee kick motor —A rocket motor that fires at the apogee of an oval transfer orbit to turn it into a circular geostationary orbit.

attitude —The position of the vehicle; for example, flying tail-first with cargo bay toward the earth.

avionics —The electronic systems and instruments to monitor and control the flight.

azimuth —The angle between a fixed point on the horizon and the direction of motion. (see ELEVATION).

blackout —The loss of radio signal (LOS) during entry caused by passage of the Orbiter through the atmosphere, creating shock waves and ionization. Also, the loss of consciousness from excessive G-FORCES.

bungee —An elastic cord used to hold equipment in place; also used for exercise on the treadmill to increase the workload and hold you in place.

cathode-ray tube (CRT) —A vacuum tube similar to a television tube, on which an electronic beam can draw pictures or write words; used to display computer output.

deorbit burn —The firing of a RETRO-ROCKET to slow the spacecraft to a speed lower than that required to maintain ORBIT. On the Orbiter, this is accomplished with the orbiter maneuvering system (OMS) engines.

dock —To join two spacecraft together in orbit.

doff —To remove (as a garment).

downlink —The communications broadcast from a spacecraft to the ground (see UPLINK).

elevon —A control surface used once the returning Shuttle has entered the atmo-phere; it acts like a combination of an aircraft elevator and aileron, controlling PITCH and ROLL.

extravehicular activity (EVA) —Work done outside the pressurized part of the spacecraft; a spacesuit is worn.

flare —A PITCH-up (nose-up) maneuver that reduces speed for landing.

fuel cell —A device that mixes oxygen and hydrogen together in a controlled process to produce electricity and pure water.

g —The symbol for the force equivalent to the acceleration of earth gravity.

g-force	—Force produced on the body by changes in velocity; measured in increments of earth gravity.
geostationary orbit or geosynchronous orbit	—An orbit 22,300 miles (35,900 kilometers) from earth, where the orbital period is 24 hours long. A spacecraft in such an orbit hovers over the equator and seems to be always in the same place in the sky. This is especially important for communication satellites.
gimbal	—An attachment with hinges or ball joints to permit movement in two or three axes. Rocket nozzles are gimbaled to allow thrust in different directions for flight control.
glideslope	—The angle at which you descend in the Orbiter or other glider with respect to the ground.
hypersonic	—Refers to speeds above MACH 5— five times the speed of sound.
inertia	—The tendency of a body at rest to say at rest and a body in motion to continue in the same direction.
knot	—One nautical mile per hour; 1 nautical mile equals 1.1 statute mile.
lithium hydroxide	—A chemical compound (LiOH) used to remove carbon dioxide from the cabin atmosphere.
Mach	—The term used to describe the speed of objects relative to the speed of sound. For example, Mach 2 is twice the speed of sound.
microgravity	—The term used to describe the apparent weightlessness and fractional g-forces produced in orbit. In orbit, you essentially fall around the earth, producing a "floating" condition.
micrometeoroids	—Meteoric particles the size of grains of salt. They are plentiful in space and potentially dangerous due to the velocities involved. The spacecraft can be shielded by a barrier equivalent to a thin sheet of aluminum.
microprocessor	—A small computer that usually performs only a few specific tasks.
multiplex-demultiplex	—To multiplex and demultiplex is to convert signals into and out of computer language. It is accomplished by electronic circuits called multiplexers-demultiplexers.
Nomex	—The brand name for a synthetic fiber used as thermal insulation.
orbit	—A balance between a body's inertia, or tendency to fly off into space, and the gravitational attraction of a central object.
partial pressure	—In a container of mixed gases (like a shuttle cabin filled with an oxygen-nitrogen mix), each gas exerts a fraction of the total pressure. This *partial pressure* is proportionate to the amount of each gas in the mixture.
pitch	—Up-down rotation of the nose of the craft (see ROLL and YAW).

pounds per square inch (psi) —A measure of pressure. The metric measure is kilopascals. 1 psi × 6.895 = 1 kilopascal

retro-fire —To fire engines in the direction of motion in order to reduce forward velocity. In orbit, this permits gravity to pull you downward.

retrorocket —A rocket that fires against the direction of motion, slowing the craft. Gravity pulls the craft downward.

rocket engine —An engine that carries its own oxidizer for combustion and, therefore, needs none from the outside.

roll —To rotate about an axis from front to back (nose to tail) of the Orbiter. To the pilot, a roll is like a cartwheel (see PITCH and YAW).

rotation —Movement of the Orbiter *around* its three principal axes producing PITCH, YAW, or ROLL.

rudder —A control surface on the vertical stabilizer (tail) to control YAW.

shirt-sleeve environment —An environment that doesn't require pressurized suits or other protective garb.

software —Computer programming or instructions.

solar cell —A device incorporating purified semiconducting metals that produce electricity when struck by sunlight.

speed brake —The rudder on the Orbiter when it is split and spread to increase atmospheric drag and slow the craft during descent and landing.

sublimation —A change of state from a solid to a gas.

telemetry —Data transmitted to earth about the Orbiter, crew and experiments.

thrust —The force created by a rocket engine.

transfer orbit —An oval shaped orbit made when changing from one nearly circular orbit to another.

translation —Movement of the Orbiter *along* its principal axis.

umbilical —A connecting cable that carries electricity and life support to an astronaut during extravehicular activity.

uplink —The broadcast to a spacecraft (see DOWNLINK).

vernier engine —A small thruster for precise adjustments in Orbiter position.

vertical stabilizer —The "tail" or the aircraft containing the rudder and the speed brake.

weightlessness —See MICROGRAVITY.

yaw —Left-right rotation of the nose of the craft (see PITCH and ROLL).

ADI	attitude direction indicator
A/G	air-to-ground
AMI	alpha/Mach indicator
APU	auxiliary power unit
ATU	audio terminal unit
AVVI	altitude/vertical velocity indicator
CSS	control stick steering
DCM	displays and controls module
DOD	Department of Defense
EMK	emergency medical kit
EMU	extravehicular mobility unit
ET	External Tank
EVA	extravehicular activity
FRCS	forward reaction control system
GMT	Greenwich mean time
GSFC	NASA Goddard Space Flight Center
HAL/S	high-order assembly language/shuttle
HSI	horizontal situation indicator
HRSI	high-temperature reusable surface insulation
IUS	inertial upper stage
IVA	intravehicular activity
LCC	Launch Control Center (KSC)
LDEF	Long-Duration Exposure Facility
LED	light-emitting diode
LOS	loss of signal
LRSI	low-temperature reusable surface insulati
MBK	medications-and-bandage kit
MCC	Mission Control Center
MCC-H	Mission Control Center at Houston
MET	mission-elapsed time
MLP	Mobile Launcher Platform
MMS	multimission modular spacecraft
MMU	manned maneuvering unit
OMS	orbital maneuvering system
OPF	Orbiter Processing Facility (KSC)
PAM	payload assist module
PLBD	payload bay door
PLSS	portable life-support system
POS	portable oxygen system
RCC	reinforced carbon-carbon
RCS	reaction control system
RHC	rotational hand controller
RMS	remote manipulator system
RSS	rotating service structure
SCA	Shuttle carrier aircraft
SCAPE	self-contained atmospheric pressure ensemble
SOMS	Shuttle Orbiter medical system
SRB	Solid Rocket Booster
SSME	Space Shuttle Main Engine
SSUS	spinning solid upper stage
STS	Space Transportation System
tacan	tactical air navigation
TAEM	terminal area energy management
TDRS	Tracking and Data Relay Satellite
TPS	thermal protection system
WCS	waste collection system

Aerojet Liquid Rocket Company Sacramento, Calif. 95813	**Orbiter maneuvering system engines**
AiResearch Manufacturing Company of California Torrance, Calif. 90509	**Air data transducer assembly and computer** **Safety valve (cabin air pressure)** **Solenoid valve (shutoff, air)**
Albany International Company FRL Dedham, Mass. 02026	**Nomex felt (for the thermal protection system)**
Ball Aerospace Systems Division Boulder, Colo. 80302	**Star tracker**
Bendix Corporation Navigation and Control Group Teterboro, N.J. 07608	**Airspeed altimeter** **Vertical velocity indicator** **Surface position indicator**
The B.F. Goodrich Company Akron, Ohio 44318	**Main and nose landing-gear wheel and main landing-gear brake assembly**
Boeing Aerospace Company Seattle, Wash. 98124	**Carrier aircraft modification**
CCI Corporation The Marquardt Company Van Nuys, Calif. 91409	**Reaction control system thrusters**
Conrac Corporation West Caldwell, N.J. 07006	**Engine interface unit (main propulsion system)**
Corning Glass Works Technical Products Division Corning, N.Y. 14830	**Windows, windshield, and side hatch window** **Glass-ceramic retainers (for TPS tiles)**
Cutler-Hammer, Inc. AIL Division Farmingdale, N.Y. 11735	**Microwave scanning-beam landing- system navigation set**
Fairchild Republic Company Farmingdale, N.Y. 11735	**Vertical tail**
General Dynamics Corporation Convair Division San Diego, Calif. 92138	**Mid fuselage**
Honeywell, Inc. Avionics Division St. Petersburg, Fla. 33733	**Flight-control system displays and controls**
Hughes Aircraft Company Space and Communications Group Los Angeles, Calif. 90009	**Ku-band radar/communication system**

Hydraulic Research Textron Valencia, Calif. 91355	Servo actuator elevon-electro command hydraulics Four-way hydraulic system flow control pressure valve
IBM Corporation Federal Systems Division Oswego, N.Y. 13827	Mass-memory and multifunction cathode- ray-tube display subsystem General-purpose computer and input-output processor
Instrument Systems Corporation Telephonics Division Huntington, N.Y. 11743	Audio distribution system (voice and tonal signals)
Lear Siegler Instrument Division Grand Rapids, Mich. 49508	Attitude direction indicator
Lockheed-California Company Burbank, Calif. 91520	Static and fatigue testing of orbiter structure
Lockheed Missiles and Space Company, Inc. Sunnyvale, Calif. 94088	High- and low-temperature reusable surface insulation
Martin Marietta Corporation New Orleans, La. 70189	External Tank
McDonnell Douglas Astronautics Company Huntington Beach, Calif. 92647	Solid Rocket Booster structure
McDonnell Douglas Corporation St. Louis, Mo. 63166	Aft propulsion system
Menasco Burbank, Calif. 91510	Main and nose landing-gear shock struts and brace assembly
Northrop Corporation Precision Products Division Norwood, Mass. 02062	Rate gyro assembly
Rockwell International Rocketdyne Division Canoga Park, Calif. 91304	Space Shuttle Main Engine
Rockwell International Space Systems Group Downey, Calif. 90241	Space Shuttle Orbiter System integration
Rockwell International Tulsa Division Tulsa, Okla. 74151	Cargo-bay doors

Spar Aerospace Limited
Toronto, Ont. M6B 3X8
Canada

Remote manipulator system

Sperry-Rand Corporation
Flight Systems Division
Phoenix, Ariz. 85036

Automatic landing
Multiplexer-demultiplexer

Sundstrand Corporation
Rockford, Ill. 61101

Auxiliary power unit
Rudder-brake actuation unit
Actuation unit (body flap)
Hydrogen recirculation pump
**   assembly (main propulsion system)**

Thiokol Corporation
Wasatch Division
Brigham City, Utah 84302

Solid Rocket Booster motors

TRW
Electronic Systems Division
Redondo Beach, Calif. 90278

S-band payload interrogator
S-bank network equipment
Network signal processor
Payload signal processor

United Technologies Corporation
Chemical Systems Division
Sunnyvale, Calif. 94088

Solid Rocket Booster separation
**   motors**
Propulsion for inertial upper stage

United Technologies Corporation
Hamilton Standard Division
Windsor Locks, Conn. 06096

Atmospheric revitalization subsystem
Freon coolant loop and flash evaporator
**   system**
Water boiler, hydraulic thermal
**   control unit**
Shuttle spacesuit

United Technologies Corporation
Power Systems Division
South Windsor, Conn. 06074

Fuel-cell powerplant

United Technologies Corporation
United Space Boosters, Inc.
Sunnyvale, Calif. 94088

Solid Rocket Booster assembly
**   (checkout, launch)**

Vought Corporation
Dallas, Tex. 75265

Leading-edge structural subsystem
**   and nosecap, reinforced carbon-carbon**
Cargo-bay door radiator and flow-
**   control assembly system**

Westinghouse Electric Corporation
Aerospace Electrical Division
Lima, Ohio 45802

Remote power controller
Electrical system inverters

Westinghouse Electric Corporation
Systems Development Division
Baltimore, Md. 21203

Master timing unit